T5-AGA-371

Science and Innovation as Strategic Tools for Industrial and Economic Growth

NATO ASI Series

Advanced Science Institutes Series

A Series presenting the results of activities sponsored by the NATO Science Committee, which aims at the dissemination of advanced scientific and technological knowledge, with a view to strengthening links between scientific communities.

The Series is published by an international board of publishers in conjunction with the NATO Scientific Affairs Division

A **Life Sciences** B **Physics**	Plenum Publishing Corporation London and New York
C **Mathematical and Physical Sciences** D **Behavioural and Social Sciences** E **Applied Sciences**	Kluwer Academic Publishers Dordrecht, Boston and London
F **Computer and Systems Sciences** G **Ecological Sciences** H **Cell Biology** I **Global Environmental Change**	Springer-Verlag Berlin, Heidelberg, New York, London, Paris and Tokyo

PARTNERSHIP SUB-SERIES

1. **Disarmament Technologies**	Kluwer Academic Publishers
2. **Environment**	Springer-Verlag / Kluwer Academic Publishers
3. **High Technology**	Kluwer Academic Publishers
4. **Science and Technology Policy**	Kluwer Academic Publishers
5. **Computer Networking**	Kluwer Academic Publishers

The Partnership Sub-Series incorporates activities undertaken in collaboration with NATO's Cooperation Partners, the countries of the CIS and Central and Eastern Europe, in Priority Areas of concern to those countries.

NATO-PCO-DATA BASE

The electronic index to the NATO ASI Series provides full bibliographical references (with keywords and/or abstracts) to more than 50000 contributions from international scientists published in all sections of the NATO ASI Series.
Access to the NATO-PCO-DATA BASE is possible in two ways:

- via online FILE 128 (NATO-PCO-DATA BASE) hosted by ESRIN, Via Galileo Galilei, I-00044 Frascati, Italy.

- via CD-ROM "NATO-PCO-DATA BASE" with user-friendly retrieval software in English, French and German (© WTV GmbH and DATAWARE Technologies Inc. 1989).

The CD-ROM can be ordered through any member of the Board of Publishers or through NATO-PCO, Overijse, Belgium.

Series 4: Science and Technology Policy – Vol. 4

Science and Innovation as Strategic Tools for Industrial and Economic Growth

edited by

Carlo Corsi
Consorzio Roma Ricerche,
Rome, Italy

Kluwer Academic Publishers

Dordrecht / Boston / London

Published in cooperation with NATO Scientific Affairs Division

Proceedings of the NATO Advanced Research Workshop on
Science and Innovation as Strategic Tools for Industrial and Economic Growth
Moscow, Russia
October 24–26, 1994

A C.I.P. Catalogue record for this book is available from the Library of Congress.

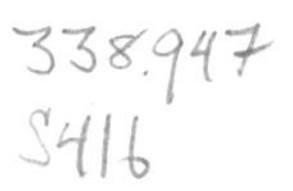

ISBN 0-7923-3903-7

Published by Kluwer Academic Publishers,
P.O. Box 17, 3300 AA Dordrecht, The Netherlands.

Kluwer Academic Publishers incorporates the publishing programmes of
D. Reidel, Martinus Nijhoff, Dr W. Junk and MTP Press.

Sold and distributed in the U.S.A. and Canada
by Kluwer Academic Publishers,
101 Philip Drive, Norwell, MA 02061, U.S.A.

In all other countries, sold and distributed
by Kluwer Academic Publishers Group,
P.O. Box 322, 3300 AH Dordrecht, The Netherlands.

Printed on acid-free paper

Printed in the Netherlands

TABLE OF CONTENTS

PREFACE ix

ACKNOWLEDGEMENTS x

ABOUT THE WORKSHOP xi

LIST OF PARTICIPANTS xvii

SMALL INNOVATIVE ENTERPRISES IN RUSSIA -
PROBLEMS AND PROSPECTIVES OF DEVELOPMENT". 1
KUDRYA S.V

"FUTURE TRENDS OF INNOVATION TECHNOLOGY
AND INDUSTRIAL DEVELOPMENT" 5
CORSI C.

INNOVATION ACTIVITY AND FIRM SIZE 15
QUINTIERI B.

MANAGING TECHNOLOGY IN LARGE MULTIBUSINESS FIRMS 21
VEREDICE G.

CULTURES OF TECHNOLOGICAL INNOVATION IN RUSSIA 35
RABKIN Y. M.1

ENTREPRENEURSHIP IN RUSSIA 51
BORTNIK I.M.

"ECONOMIC REFORM IN RUSSIA AND GOVERNMENT
REGULATIONS OF SCIENCE, TECHNOLOGY
AND ACTIVITIES OF R&D SMES 57
FEDOROVICH V.A.

BRINGING RESEARCH AND INNOVATION TO THE MARKETS 61

DEL CANUTO U.

INNOVATION IN CRISIS:
HUNGARY BEFORE AND AFTER THE WATERSHED OF 1989 71
SZÁNTÓ B.

THE ROLE OF SCIENCE PARKS IN THE PROCESS
OF INNOVATION 85
PARRY M.

ASSOCIATION FOR THE PROMOTION OF SMALL INNOVATION ENTERPRISES,
TECHNOLOGY CENTRES AND TECHNOPOLISES,
MOSCOW - RUSSIA 97
CORTI E.

SCIENCE AND TECHNOLOGY STATISTICS IN RUSSIA:
TRANSFORMATION IN LINE WITH THE INTERNATIONAL
STANDARDS 125
MINDELI L.

FOSTERING A RUSSIAN HIGH TECHNOLOGY INDUSTRY
IN AN OPEN MARKET ENVIRONMENT 135
COLETTI A.

R&D ROLE FOR THE IMPROVEMENT OF SMES INNOVATION,
COMPETITIVENESS, AND CONTRIBUTION TO
SOCIO-ECONOMIC GROWTH 143
NEGRI L.

CONSIDERATIONS ON REALIZATION OF TECHNOLOGY
TRANSFER IN RUSSIA 149
GORAK A.

INNOVATION AT BRUNEL SCIENCE PARK 153
RUSSELL P.

COMMENTS ON WORKSHOP PAPERS 157
TURNER J.R.

DYNAMICS OF INNOVATION 159
SOARES O. D. D.

"OUTBOUND TRANSFERS OF TECHNOLOGY IN A COOPERATIVE
PERSPECTIVE" 165
SZANTO B.

INDEX 169

PREFACE

The great, complex and rapid change happening in former Soviet Union is overfloading an impressive impact on the western world, especially Europe, and, in the close future, on the global world.
Most of this change is generating positive effects and even more optimistic expectations, but surely the difficulties to support and to render these results real and longlasting cannot be underestimated .
In fact, difficulties in the adaptation, especially of the most important Countries capabilities, like R&D process and Innovation development and transfer, are being evidenced in the transitional period to completely new socio-economic and political conditions.
For the above reasons various Conferences and Meetings have been organised on international base, most of them taking care of identifying and developing recommendations for improving organisation of Science in East Europe and reshaping the research in Science and Technology in the context of new socio-economic conditions.
These efforts were mainly confined to scientific research that was considered one of the most important wealth's of Soviet Union, giving not specific attention to the strategic importance of Science and, even more, Innovation for industrial and socio-economic growth in the new N.I.S. Countries.
Furthermore, the impressive speed of change in Innovation on the global market coupled to the enormous change realised by N.I.S. Countries, especially by the leader Russia, is accelerating the need of an operating solution capable of linking these Countries with the Western World, rules and market, starting from Europe.
So, Science and Innovation are to be treated and organised as strategic tools for industrial and economic growth and have to act and operate synergystically within the global world scenario, possibly using the networks operating in the western world and their rules, methodologies and, even more important, their quality specifications.
For these reasons, and probably for much more important reasons which can embrace vital human values like Democracy and Friendship, our proposal to support this Symposium was approved and realised as NATO ASI.
Having mentioned Friendship, it is dutiful and rightful to dedicate the closing note to remember Dr. Ygor Valaiev, the best example of Friendship, sheltered by an unpenetrable Russian language and a solid massive figure, a Man capable of creating friendship and cooperation by shaking strongly the hands and looking deeply in the eyes, who suddenly left us before seeing the growth of this strategy based on friendly cooperation for achieving future successful socio-economic results for His Country : "ROSSIA".

ACKNOWLEDGEMENTS

A.M.I.P. and ROMA RICERCHE on behalf of all participants who attended the Advanced Research Workshop NATO International Scientific Exchange Programmes would like to express deep gratitude to the NATO Scientific Affairs Division and its Office Head Dr.Alan Jubier for the financial support and programme suggestion.
In addition thanks are due to the Russian colleagues for their unvaluableand unceasing support and interest in the meeting organisation, in particular is to be underlined the outstanding engagement of Dr.Sergey Kudria, President of AMIP and Co-Director of the workshop and of Dr Boris Roziskiy, Chief Project for Russian organisation, with his AMIP's colleagues; special thanks are due to Dr.ssa Claudia Cardone, Chief Project for the Nato Countries, who has done an outstanding job in organisation and friendly guesting of the meeting with a charming mixture of deep Russian and Italian culture and a patient and unyielding work in supporting the Proceedings editing.

Foreword

The NATO Advanced Research Workshop "Science and Innovation as Strategic Tools for Industrial and Economic Growth" was held in Moscow on 24 - 26 October 1994.
The Advanced Research Workshop (ARW) was organized under the decision of the NATO Assistant Secretary General for Scientific and Environmental Affairs on the recommendation of the Advisory Panel on the Priority Area on Science and Technology Policy / Human Resources.
The decision was taken in line with the NATO programme of humanitarian contacts with CIS countries and countries of Eastern and Central Europe.
The Workshop was organized under financial support of the NATO Scientific Committee with participation of the Ministry for Science and Technological Policy of the Russian Federation and the Russian Academy of Sciences.
The principal organizers of the Advanced Research Workshop:
Association for the Promotion of Small Innovation Enterprises, Technology Centres and Technopolises (AMIP), Russia and Consorzio Roma Ricerche (CRR), Italy.
The Workshop was attended by 55 representatives of NATO ember-countries and NATO countries-partners. The list of participants is enclosed.

ABOUT THE WORKSHOP

Aims of the workshop

The Workshop had the following principal aims:

- Promotion of cooperation in the area of innovation activities between Western countries and countries of Eastern and Central Europe;
- Analysis of experience, tendencies and problems of development of small and medium-sized entreprises in the field of innovation and relative governmental policies;
- Analysis of efficiency of the Western assistance to Eastern and Central European countries, especially to Russia and other CIS countries, in innovation technology transfer and dissemination, in partnership in research and development, production, marketing, etc;
- Working out recommendations on innovation development policies, and relative practical steps for the countries which economies are in process of reforms, paying special attention to the development of the sector of small and medium-sized innovative enterprises.

The specified aims and questions of particular attention of the Workshop are given in the Justification of the Workshop.

Programme Organization

The Organizing Committee of the ARW was headed by:

1. Dr. Sergey V.Kudrya, Chairman of the Board, General Director of the Association for Promotion of Small Innovation Enterprises, Technology Centres and Technopolises (AMIP), RUSSIA;
2. Prof. Carlo Corsi, General Director of Consortium Roma Ricerche, ITALY;
3. Prof. Nicolay P.Laverov, Vice-President of Russian Academy of Science, RUSSIA.
4. Chief Programme for Western Europe: Dott.ssa Claudia Cardone, Consortium Roma Ricerche
5. Chief Programme for Eastern Europe: Dr. Boris Rozinsky

The main reports at the Workshop were done by 16 Rapporteurs:

1. Dr. Kudrya S.V. General Manager of AMIP, (Russia) " Small Innovative Enterprises in Russia - problems and prospectives of development".
2. Member of the Academy Laverov N.P., Vice-President of the Russian Academy of Sciences (Russia)
3. Dr. Yakobashvili Z.A., Deputy Minister for Science and Technological Policy (Russia) "International Scientific and Technological Cooperation as one of the Main Tools for Integration of Russia in the World Economic Development".
4. Prof. Corsi C., General Manager of Consortium Roma Ricerche (Italy) "Future

trends of Innovation Technology and Industrial Development".

5. Prof. Quintiery B., Faculty of Economy and Business Administration - Rome University "Tor Vergata" (Italy) "Finance, Technological Change and Industrial Growth with Reference to SMEs".
6. Prof. Veredice G., R&D V. Director, Finmeccanica (Italy) "Managing technology in large multibusiness firms".
7. Prof. Fedorovich V.A., Head of the Division of General Problems of American Economy, Institute of the US and Canada Studies of the Russian Academy of Sciences (Russia) "Economic reform in Russia and Government regulations of Science, Technology and Activities of R&D SMEs".
8. Dr. Gumen R.G., Director of the UNIDO Centre in Moscow "The UNIDO activities in Technology transfer and Promotion of Investments in Small and Medium sized Enterprises".
9. Dr. Del Canuto U., President IRI Management (Italy). "Privatization process: the Italian Experience"
10. Prof. Rabkin Y.M., President of Periscience Inc., (Canada) "Outbound Transfers of Technology in a Cooperative Perspective".
11. Dr. Szanto B., Budapest Technical University, (Hungary) "Crisis of Innovations in Hungary: before and after 1989"
12. Dr. Parry M., General Manager of the Surrey University (UK) "Science Parks - their Role in Innovation Development".
13. Prof. Corti E., Vice-President of the Technologies for International Technoparks, (Italy) "Technology Diffusion and Transfer in the Eastern Countries".
14. Prof. Bortnik I.M., General Director of the Fund for the Promotion of the Development of Small Forms of Entrepreneurship in S&T, (Russia)
15. Prof. Kozlov G.V., Deputy Minister for Science and Technological Policy (Russia) "Development of innovation entrepreneurship in Russia".
16. Prof. Mindeli L.E., Director of the Centre of Science and Research of Statistics (Russia) "Science and Technology Statistics in Russia. Transformation in line with the International Standards"

Meeting Objectives

Science and Technology are the most strategic tools in advanced countries for high level economy and industry.
The scenario of Science and Technology in Russia shows very high-level scientific know-how, but a great difficulty to transfer this know-how to industrial applications and, as consequence, it shows very weak and gracile industrial structure. The only exception is the military industry, where advanced perfomance and high level of products have been achieved. The above considerations push to start an urgent action which can tackle the important task of utilizing the advanced level of research into industrial production by making a possible use of military conversion.
The above problems (especially the military conversion) are difficult tasks to be achieved also in Western Countries, but in the case of Russian industries this is

worsened by the difficulty to enter to the international market due to the missing of a kind of an International Technological Language that is based on common rules for the processes of production and maintenability (e.g. certification, quality control, CAD/CAM/CIM).
The complexity of the problem suggests the necessity to identify to specific items, very well defined, which would have anyway economic dimension and could be assumed as examples to tackle the problem.

Meeting Summary Report

The participants of the Workshop considered a wide range of important problems in the field of innovation activities having basic significance for providing high level of national industrial and economic development and for expanding international scientific and technological cooperation with active participation of CIS countries, Central and Western Europe.
Special emphasis was made on the questions related to the state regulations in science, technology and activities of small and medium sized innovative enterprises and to the creation for this purpose of the appropriate infrastructures.
The participants of the Workshop stressed the necessity for elaboration and application in Russia of the proper legislation in science and technology area and on activities of small innovative enterprises, in particular, in view of a transitive period of the Russian economy and with account of the experience of the industrially developed countries of the West.
There were specifically stressed the importance of development and observance of the norms of legal protection of the intellectual property rights in the countries in the process of reformation.
The areas of advanced scientific and technological research, which could serve examples of scientific know-how for transfer to industry (space, energy, environment and others), were subjected to a detailed analysis and identification.
The participants underlined the significance and complexity of privatization process in innovation area.
Special attention in the statements of participants was paid to the necessity of creation of Common International Technology Information Networks and International Technology Transfer Centres, that would require the achievement of high level of applied technologies, like multifunctional computerized flexible technologies, technologies CAD/CAM/CIM and others. The creation of such structures should promote mutual dissemination and exchange of the most advanced modern technologies, available in different countries.
The Workshop stressed the necessity of carrying out a gradual conversion process of defense industries, and also underlined a special and very important role in this process of small and medium sized innovative enterprises - the carriers and creators of the most updated and advanced technologies, best-selling in international market. There was underlined the necessity of urgent measures, aimed at practical application in industrial

production of the achieved scientific results (first of all coming from the military industries).
The reports, made at the Workshop, reflected the experience of technology exchange between the CIS countries and the countries of Europe, USA and Canada. Besides, there were pointed out the difficulties, arising in the process of technology transfer, among which still existing in this sphere the barriers and restrictions, particularly at the state level, were mentioned. The proposals for their elimination were made.
The experience of international organizations in the area of technology transfer was considered by examples of UNIDO activities and of a number of international technoparks.

Final conclusions and recommendations

The participants of the Workshop worked out the following recommendations for creation of necessary conditions for comprehensive industrial and economic development:

Government regulations in the sphere of science and technology.

The participants of the Workshop are firmly convinced, that for the successful continuation of the process of economic reforms in Russia, CIS countries and Central Europe, for the effective development of science and technology and for the creation of equal in rights conditions for technological cooperation, these countries should study and apply the best of the economic experience, generally accepted in the world, with account to specific features of every particular country. With this aim they should consistently follow the following measures:

- Carry out deep analysis of innovation activities of advanced countries at present, as well as for the past 30-40 years, and of the appropriate models of innovation development;
- Develop and introduce a relevant mechanizm of state regulations in science, technology and economy including a complex system of state support for small innovative entrepreneurship;
- Put into operation a dynamic market mechanizm, including competitive state market of R&D products;
- Consecutively replace an administative and command system of planning by a government contract system (including practice of concluding government contracts in the sphere of research and development), to be considered as a main economic tool of commercial, industrial and technological cooperation between countries;
- Create an effective system of protection of intelligent property rights of the partners in international technological cooperation.

Creation of International Information Networks and Centres for exchange and dissemination of innovations.

As a primary task to consider is the creation of Integrated Networks of Innovation Centres for technology and know-how transfer with the aim of effective exchange of technological information that would use the existing and newly created International Information Networks.
With the purpose to expand international cooperation with the countries in the process of reforms in innovative sphere there should be created a common international technological language, based on common rules and norms regulating R&D activities, production and realization of goods, and mutual exchange of technologies and know-how. That will promote the prompt entry of these countries into the world market.
The participants of a Workshop recognized the necessity to establish an International Centre for Technology Transfer between Russia and Western countries, which activities would be directed, in particular, on creation of a information infrastructure for small and medium sized innovative enterprises.
In accordance with the preliminary agreements between the Consortium " Roma Ricerche ", the Association "AMIP" and the Ministry for Science and Technological Policy of the Russian Federation there was adopted a recommendation to apply to the Scientific Committee of NATO to consider a possibility of assistance in the implementation of the project on creation of a Joint CAD/ CAM /CIM Centre for small and medium sized enterprises of Russia.

Legal protection of activities of small and medium sized innovative enterprises in Russia

The participants of the Workshop confirmed necessity of working out and application in Russia of apppropriate legislation, ensuring necessary conditions for development of science and technology, in particular with respect to the sector of small and medium sized enterprises with account of the present state of the Russian economy and international experience.
The necessity of the prompt adoption in Russia of the law on state support for small and medium sized enterprises, especially for those acting in innovation sphere, was especially underlined.

Privatization.

Alongside with the development of privatization process in the field of science and technology, that, in opinion of the participants of the Workshop, should speed up the process of development of small business and have favorable effect on solving the existing problems of countries in the process of reforms into the world market system, there should be taken into account a strategic role of the state in defining national priorities of development in S&T area.

Transfer of technologies from R&D sphere to industry.

The participants of the Workshop consided expedient to carry out a comprehensive analysis of methods for transfer of advanced technological developments and know-how from defense industries to civil applications with account of international norms and standards in the field of quality control and certification that is necessary for entering the international market, taking into special consideration the development of sector of small and medium sized enterprises.

International transfer of technologies.

The participants recognized vital importance and necessity of easing and removing the existing restrictions, in particular at the state level, hindering free exchange of technologies and R&D results.

Personnel training.

The participants of the Workshop consided it essential to widen the existing system for experts training for the countries of Eastern and Central Europe especially in modern methods of quality control and innovation management.
Taking into account actuality and importance of questions discussed at the Workshop, the participants considered expedient the organization on a regular basis of international conferences and advanced research workshops on innovation, technologies and cooperation of small and medium enterprises.
With this regard the participants recommended to the Organizing Committee of the Workshop to work out a list of relevant subjects and submit it to the NATO Scientific Committee for consideration of possibility of organization in 1995 of a new meeting of specialists.
There should be forseen necessary support from the national governmental and other organizations involved in the process of innovation development.

LIST OF PARTICIPANTS

1.	Asche	Michael	Universitat Dortmund Transferstelle D-44221 Dortmund	GERMANY
2.	Belaev	Spartac	"Kurchatov Institute" of Nuclear Energy of Moscow, Kurchatov Square,46123182 Moscow	RUSSIA
3.	Bohm	Janos	INMAS Mechatronica Director Budapest XIIITeve utca1/b-c	HUNGARY
4.	Bortnik	Ivan M.	Fund for the Promotion of the Development of Small Forms of Entrepreneurship in S&T General Director	RUSSIA
5.	Cardone	Claudia	Consorzio Roma Ricerche Centro per l'Innovazione Via Orazio Raimondo,8 00173 Rome	ITALY
6.	Coletti	Alessandro	Italian Trade Commission 28 Webster Avenue Harrison New York 10528	USA
7.	Corsi	Carlo	Consorzio Roma Ricerche, Salita di San Nicola da Tolentino, 1/b 00187 Rome	ITALY
8.	Corti	Eugenio	Technologies for International TechnoparksSt.Petersburg Deputi General Director LARA Engineering s.r.l. 5204 S Maria Formoza 30122 Venice	ITALY
9.	Courregelongue	Gilles	CEO of Defence Counceil International Manager Paris	FRANCE
10.	Daukeev	Dias K.	Ministry of Science and New Technologies Deputy Minister Alma-ataMasanchi57	KAZAKHSTAN
11.	Del Canuto	Umberto	IRI Management Via Piemonte, 60 00187 Rome	ITALY
12.	Egorov	Sergey A.	State Innovation Found of Ukraine Lvovskaya Square, 8 254655, Kiev - 53	UKRAINE
13.	Fedorovich	Vladimir A.	Institute of the US and Canada Studies of the Russian Academy of Sciences	RUSSIA
14.	Fonotov	Andrey G.	Ministry of Scence and Technological Policy of Russia 11, Tverskaya 103905, Moscow	RUSSIA

15.	G.Gumen	Robert	UNIDO Centre in Moscow General Director	RUSSIA
16.	Gorak	Andrzej	Universitat Dortmund, Department of Chemical Engeneering Chair of Thermal Process Engineering D-44221 Dortmund	GERMANY
17.	Goulet	Julie	Department of History McGill University 855 Sherbrooke Street West Montréal, PQ H3A 2T7	CANADA
18.	Jamankyzov	Nasytbek K.	State Committee on Science and New Technologies Vice-Chairman 720000BishkekTyinoystanova257	KYRGYZSTAN
19.	Janosh	Böhm	J. V. Inmash Mechatronica, KFKI Economy Chamber of Hungary Budapest XIII, Teve n. 1/b -C 1365 Budapest 5	HUNGARY
20.	Jumaliev	Kubanizshbe k M.	Ministry of Education and Science Tyinystanova, 257 720000, Bishkek	KYRGYZSTAN
21.	Karapetyam	Suren A.	Sociological Department, Institute of Economy Movcesa Khorenatsi, 13 375010, Yerevan	ARMENIA
22.	Kolesnikov	L.A.	Ulyanovsk State Technical University	RUSSIA
23.	Kolyakin	V.N.	Ministry for Science and Technological Policy Councellor	RUSSIA
24.	Kovalchuk	Vasily B.	State Innovation Fund of Ukraine Lvovskaya Square, 8 254655, Kiev - 53	UKRAINE
25.	Kozlov	Gennadiy V.	Ministry for Science and Technological Policy of the Russian Federation Deputy Minister 103905Moscow Tverskaya street11	RUSSIA
26.	Kroo	Norbert	Institute of Solid State Phisics, KFKI M.T.A. Budapest XII, Konkoly Thege Ut Budapest	HUNGARY
27.	Kudaybergenov	Sarkyt E.	National Academy of Sciences Shevchenko, 28 480021, Alma - Ata	KAZAKHSTAN
28.	Kudrya	Sergey V.	AMIP 11, Tverskaya 103905, Moscow	RUSSIA
29.	Lavërov	Nicolay P.	Russian Academy of Sciences, Leninsky avenue, 14 - 117901, §Moscow	RUSSIA
30.	Letailleur	Thierry	CEO of Defence Counceil	FRANCE

			International Paris	
31.	Minaev	E.S.	Moscow Aviation Institute	RUSSIA
32.	Mindeli	Levan. A.	Centre for Science Research and Statistics of Russia. 11, Tverskaya-103905, Moscow	RUSSIA
33.	Negri	Lionello	UTIBNoT Via Tiburtina 770 00159 Rome	ITALY
34.	Oehler	Birgitta	Universitat Dortmund, Transferstelle, D-44221 Dortmund	GERMANY
35.	Orlov	Leonid P.	Science Department, Council of Ministers of Byelorussia The House of Governament 220050, Minsk	BYELORUSSIA
36.	Parry	Malcom	The Surrey Research Park Guildford 30 Frederick Sanger Road Surrey GU2 5YD	UK
37.	Prozorov	D.E.	Moscow Aviation Institute	RUSSIA
38.	Quintieri	Beniamino	Economy and Business Administration University "Tor Vergata" Via Orazio Raimondo 8 Rome	ITALY
39.	Rabkin	Yakov M	Periscience Inc., 6229 Deacon Montréal, Qc H3S 2P6	CANADA
40.	Rastas	C.	Global Development Services Inc. Represenrative in Moscow	USA
41.	Rhéaume	Charles	Department of History McGill University 855 Sherbrooke Street West Montréal, PQ H3A2T7	CANADA
42.	Richard	Frederic	UNIDO Small and Medium Enterprises Dept.,	UNIDO
43.	Rozinski	Boris	AMIP 11, Tverskaya 103905, Moscow	
44.	Russell	Peter F.N.	Brunel Science Park, Cleveland Road, Uxbridge, Middlesex UB8 3PH	UK
45.	Safonov	Lev M.	Moscow Chamber of Real Estate Manager	RUSSIA
46.	Sazonov	B.V.	Institute of System Analysis Russian Academy of Sciences	RUSSIA
47.	Sillard	Ives	CEO of Defence Council International Paris	FRANCE
48.	Soares	Olivério D.D.	CETO Lab. De Fisica - Faculdade de Ciencias Porto University Prç Gomes Teixeira 4000 Porto	PORTUGAL

49.	Szanto	Borisz	J. V. Inmash Mechatronica, KFKI Economy Chamber of Hungary Budapest XIII, Teve n. 1/b -C 1365 Budapest 5	HUNGARY
50.	Turner	J.R.	Unit 1A Mountjoy Research Centre, Stockton Road Durham DdH1 3SW	UK
51.	Ukuev	Batyr T.	Research Institute of Economy, National Academy of Sciences Leninsky Avenue, 265-A 720071, Bishkek	KYRGYZSTAN
52.	Vasin	V.A.	B.Sc. Centre of Science Research and Statistics Researcher Moscow	RUSSIA
53.	Veredice	Giuseppe	Finmeccanica, Viale Pilsudsky,92 Rome	ITALY
54.	Voskoboy	Alexey S.	Russian Union of Industrialists and Entrepreneurs The Old Square, 10/4 103070, Moscow	RUSSIA
55.	Yakobashvili	Zurab A.	Ministry of Scence and Technological Policy of Russia 11, Tverskaya 103905, Moscow	RUSSIA

"SMALL INNOVATIVE ENTERPRISES IN RUSSIA - PROBLEMS AND PROSPECTIVES OF DEVELOPMENT".

KUDRYA S.V
General Manager
of AMIP, Moscow, Russia

Today work of the NATO international SEMINAR "Science And Innovations - strategic tools of industrial and economic growth " begins. Urgency of the question delivered for discussion cause no doubts: it largely, to my mind, explains wide international representation of its participants.

Nowadays we've got representatives of thirteen countries at the Seminar. Among them both representatives of NATO countries as well as those from former socialist ones. On behalf of Organizing Committee of the Seminar allow me to thank those all present for attention, displayed to our event, for the consent to participate in.

The idea of realization of the Seminar has arisen during long-term cooperation of the main organizers of a given event: Russian Association for promotion of small innovation enterprises, technological centres and technopolises and Italian Consortium "Roma Ricerche".

As the Chairman of the Board of Association I would like in a few words to tell you about main directions of its activity. The Association is called to render assistance to small technological business, with the purposes of creation conditions for its effective activity. The Association promotes application of progressive scientific ideas R/S and technologies to production; assists in participation of SME's in the state technological programs; supports Russian innovative enterprise in realization of international relations, with the aims of creation of joint ventures, manufactures, technical centres, in search of the partners in other countries.

The identity of the tasks and problems, put before Association and Consortium "Roma Ricerche", oriented on innovation progress and development was the pledge of their successful cooperation.

A year ago the First International Conference "Innovations, Technologies and Cooperation between SME's of a EC and Russia " was organized by joint efforts and successfully carried out in Moscow. Preparation of a following conference is being conducted at present, on which it is supposed to continue exchange of opinions on a similar theme.

The Present Seminar is supported by the Scientific Committee of the NATO and Ministry of a science and technical policy of Russia.

Even some years ago it was impossible to imagine cooperation between these organizations. This is a comforting fact, that today's questions of international

C. Corsi (ed.), Science and Innovation as Strategic Tools for Industrial and Economic Growth, 1–3.

technological cooperation unite even former antipodes.
The fact, that our contacts can and should become constant, testifies the declaration of the new General Secretary of the NATO, made on the occasion of taking the office on October 17, 1994 He has said, that for the NATO now "the main task is the work with new partners on East ": the block is going to develop closer contacts with the countries of Eastern Europe equal and active participants of the programm "Partnership for the sake of peace". The development relations with Russia should become, in his opinion, " the task number two ", because without Russia NATO an not ensure "the reliable and lasting peace in Europe ".
The organizers of a Seminar hope, that its participants - representatives of scientific circles of many countries, international organizations will have an opportunity to exchange experience and ideas in the area of science and innovation processes, of technology transfer, development formation of SMEs in Russia.
However, I consider it necessary to note the following. At present the number of various programms of assistance to Russia, including technological, investment and others, are being implemented. Unfortunately the majority of those programms proved to be uneffective.
The stabilization of economic situation in Russia, development of SMEs, including technological sphere, could help foreign investments, but the conditions for reception and insignificant volume do not meet the requirements of present time and can not satisfy the Russian businessmen. In opinion of international financial establishments, the main reason of it is the fact, that the climate for investments in Russia has worsened for the last 5-6 years. For example, the financial journal " European money " has placed Russia in March 1993 on 149 place (from 169 countries) on reliability of investments. Besides difficulties of the legislative code, the foreign investors collide with many organizational and information problems. Such "internal" Russian problems should be sottled by Russians themselves.
But, on the other hand, main part of financial means, selected within the framework of programms of support of Russian enterprises, remains at the command of western firms, which finance realization of research work or, at the best, will organize training for Russian businessmen. Such practice can be considered as a loan to Russia, but not help. In any case, Russian innovative enterprises live from bad to worse.
I wish that our work on the Seminar has not become similarity of such programms, that's why I think it necessary, that in the reports, of the participants we could find not only generalizations of experience of technological development in various countries, but concrete practical recommendations for use of this experience in Russia. I hope also, that in discussions, discussion "at a round table" the participants of a Seminar will state opinion on possible directions and forms of technological cooperation of countries and organizations with Russia.
The specific point of the settlement of Russian economic problems, in search of ways out from economic crisis take formation and development of activity of small and medium sized innovative enterprises. The interrogations of the Russian businessmen show significant interest from their part to experience of foreign countries in the field of development of small business.

The transition from super monopolized state system to the market economy in Russia is very complicated. It is difficult to evaluate scales of development of small business for our country. According to statistics, in Russia it was registered about 950 thousand new small enterprise. In May 1993, in his report named 620-630 thousand small enterprises of the different forms of the property. About 8 % of the registered small enterprises are engaged in scientific activity. By evaluations, for 1993 number of SMEs was increased by 1,6.
The world practice consideres, that the small enterprise can be created in account on 30-50 persons of the population. So Russian potential of small business is about 3-5 mh enterprises.
Units.
On October 19 State Duma started considering the draft of the Law about State support of small interstate enterprises, the main aim of the Law is formation of necessary legal basis of further development of competitive market relations, demonopolisation of economy, increase of business activity of population.
So we hope, that necessary conditions for activity of innovative small enterprises will be created at last and opportunity to realize recommendation and decision, taken at the Seminar, will appear.
In summary I would like once again to express a wish, that as a result of work of the Seminar the approaches of various countries and international organizations to the decision of problems common for all countries were determined, the most effective ways of cooperation in interests of industrial and economic development are found.

"FUTURE TRENDS OF INNOVATION TECHNOLOGY AND INDUSTRIAL DEVELOPMENT"

CARLO CORSI
CONSORZIO ROMA RICERCHE
Salita San Nicola da Tolentino 1/B,
00187 Rome, Italy

1. INTRODUCTION

Coordination and links among international, national and regional initiatives for promoting and developing Innovation Centres and Technology Transfer Programmes to support economic and industrial growth is starting to be one of the key factors for future economic success.
In fact that the success of national economies are today strongly influenced by the capability of managing the innovation change, and, even more, it is of operating synergistically within networks capable of managing the rapid change and transformations of the socio-economic context, generated by the fast growth of scientific and technological know-how.
This progress is so fast and so wide (the innovative wave is investing practically all operational and functional areas: research and development, planning, finance, production, marketing and even "image"), that for an isolated industrial structure it is almost impossible to react promptly and adequately to this change to survive economically.
To answer to this basic requirement, the advanced industrial system was looking for new actions to support innovation demand, especially enlarging external cooperation with different sources of knowledge, mainly Universities and Research Centres and by creating infrastructures and organisations, finalised to support the linking between the knowledge generators (Universities and Government Research Centres) and the users (industrial and economic systems).
These structures and infrastructures, called either Science Parks or Innovation Centres or Technological Poles, depending on their size, function and behaviour, are based on the principal concept, that to operate nowadays at international level in advanced technology, it is needed to cooperate and to act synergistically for technology transfer either on local areas as well as on international base.
So, the Innovation Centres should act not only as linking of scientific organisations with local structures (Chambers of Commerce, Industrial Associations, Regional Organisations), but even more as poles of a Network to foster transfer of the scientific

C. Corsi (ed.), Science and Innovation as Strategic Tools for Industrial and Economic Growth, 5-13.

knowledge and innovation technology.
So, a diffused Network of Innovation Technology Transfer, which strongly operates in the training, is more and more growing on European base, creating an Integrated Systems Network, clustered around the more important and active Science and Innovation Centres.
The main action on international base of this Innovation Technology Transfer and Training Centres Network (ITTTC) is emerging as a capability of linking different economic and cultural structures by creating a kind of a "Common Technological Language", which can be used as a tool of synergistic cooperation and interactive exchange among various Countries.
This cooperative action, especially on science side, is going on for long time within EC, and various CEC Programmes are supporting more and more the Technology Transfer by creating Networks of various structures in different scientific and socio-economic fields.
These networks should be structured as an Integrated System of Innovation Centres, which could be one of the best way to activate and sustain cooperation and technology transfer to the Eastern European Countries, which, emerging from a closed system, completely torn away and isolated from Western advanced technogical economy, needs a tool for a friendly cooperation and, therefore, a Common Technological Language to be used as for an easy interaction instrument.

2. THE EASTERN EUROPE SCENARIO.

The Russian Federation N.I.S. is currently engaged in a vast process of transformation in which science and technology are among the main assets.
The Soviet Union's scientific capability (70% in Russia) shows human resources equal to the major OCDE areas. Even if this capability must be considerably reduced for the reorientation from military to civilian applications, it is a resource of immense importance, next to the natural resources, for the future of the N.I.S.
Although the high level of scientific know existing in NIS countries, the Eastern Europe is revealing a clear dichotomy between the outstanding scientific levels and the gracile industrial economy especially in high-tech manufacturing.
There are many reasons for explaining this evidence and some of the main problems are listed in the following:

- Institutional and financial framework in which goverment policy is developed.
- Reorganization of science structures in a confused and weak
- Need to move rapidly within a confused and complex
- Innovation scenario and climate in the difficult transition to a market economy.
- Graduate shift from military to civilian application of science and technology within the difficulty of a lack of solid programming and in presence of an economic crysis.
- Amplification of problems for I.T. development in absence of capital investment, specialized labour and product markets.
- Temptation to mantain a State involvment in presence of an excessively aggressive run to "easy gain of money"

Lack of synergistic interaction capability among the scientific culture, the innovative technologies applied to production and, very important, the market analysis.

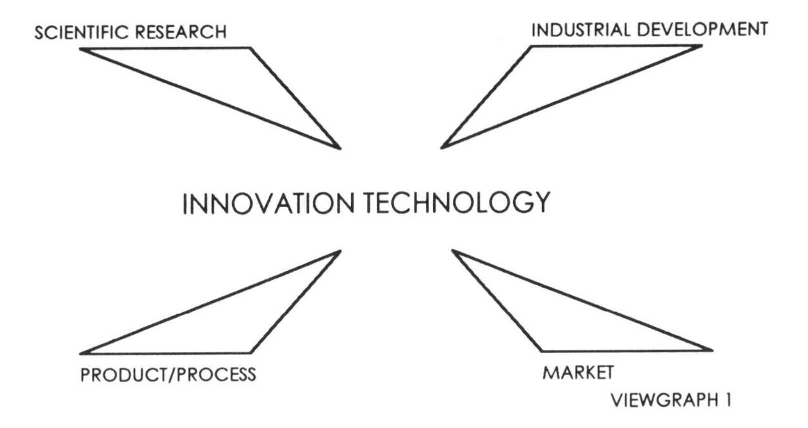

Fig. 1

The above scheme, although strongly simplified evidentiates the importance of synergist links among the four aspects and especially the link between market (including market demand) and buyers possibilities and industrial production (including quality control and products distribution and selling).
Moreover, the general definition of innovation, automatically refers to something new for someone and somewhere, introducing two other fundamental aspects of innovation technology which are to be taken into accounts: the first one reflects the international character of "space" and the other one the local character of "time".

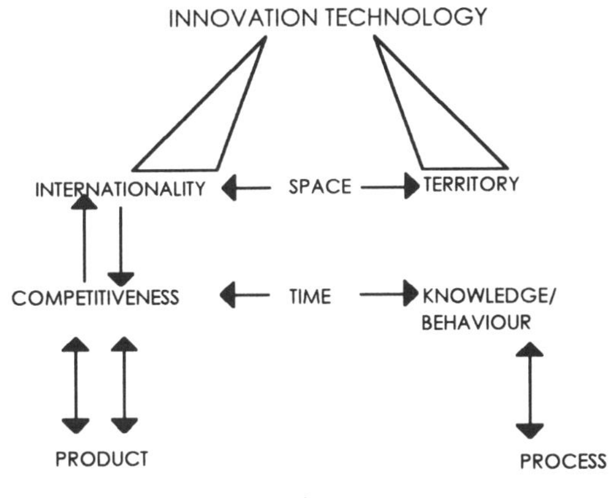

Fig. 2

These temporal and spatial aspects are rendered more complex in their interaction by the new globalization of markets, which is provoking furthers difficulties to countries, like the ex Soviet Union, which operating up to now within a closed system are

impacting problems with the free market suddenly and in a very wide spectrum problems.
These problems are moreover exploding for small organisations like SMEs, which have to counter with a new governmental culture while competing in international market.
Infact, while for great enterprises there is different situation for high tech companies linked to military market (aerospace and defence) which, if connected with international organisations, can become more and more competitive and other industrial structures which are obsolete in know-how and productivity, various problems are emerging for Russian SMEs.
First of all there is scarce knowledge in the enterpreneurships especially for market analysis and cost evaluation, then there is a diffused confusion between the role of commercialisation and the industrial production capability with the consequence of a weak link between the two activities.
Second, but non secondary, althought a strong interaction, almost a dependence from the Government Research Labs, is clearely evidencing of a high level of scientific know-how especially in the field of physics and chemistry, there is a a general lack of an adeguate know-how in quality control and service mantainance.
The above limits, especially the lackness of "Total Quality" culture for reliable exportable goods and service (maintenance reliability) and a weak industrial technology management culture in transferring high-tech pilot plant and prototyping to industrial production are strongly limiting the "great" growth possibilities of an industrial system competitive on the international free market.
On the other hand, the high level of scientific know-how, the high-tech know-how development in military and aerospace fields, the advanced prototyping, directly from R&D, coupled to a low-price, high skilness workers and good technologies in specific strategic fields (metallurgy, mechanics, optics and biotechnology) are involving future positive growth of NIS industrial technology and consequently economy.
Limiting our analysis to technological field avoiding to deal with organisational limitation evident in NIS countries like slowness and delay in decisional process, infrastructure disorganisation and other important coustraints, it is more and more evident that the link to free market economy is strongly based on NIS's capability to link the international technology level of know-how which is deeply and widely spread within the best production systems.

3. COMMON TECHNOLOGICAL LANGUAGE: TRANSVERSAL PERVASIVE TECHNOLOGIES.

The great revolution of the open market emerging in the NIS Countries is therefore more and more evidencing the need of high-tech SMEs to recover this lack and gap between Eastern and Western Europe economies.
The reasons for doing well and quickly are many and the ways to be followed can be various, but probably the most efficient and prompt ones should be based on the use the high level of human and instrumental resources existing in the NIS Countries and by introducing and/or creating the needed infrastructures (cultural and instrumental) to

support and organize these resources.
So, while it is important to use the R&D Capabilities of NIS Countries for developing deep know-how in vertical technologies in the most strategic fields (Microelectronics/ Optoelectronics, Biotechnology and New Materials), in parallel it is important to create infrastructures (Cultural and Instrumental) for developing programme cooperation in transversal and pervasive technologies (CAD/CAM/CIM, Technology Information and Advanced Processing Technologies).
Moreover, we should take in consideration that the commercial spirit of NIS peoples pushing them mainly to commercial fields, selling and buying consumer goods, strongly limiting or at least delaying the capabilities as SMEs manufacturing.
These cultural constraints are deriving also from the fact the SMEs in NIS Countries are missing adequate technological infrastructures, which till now have been developed only for supporting great governmental industries.
Moreover SMEs in NIS Countries are usually originated by high skill scientists and/or technical directors of great governmental industries, therefore people not accustomed to organize the developing and manufacturing with small and limited infrastructures; this is evidencing limitations and constraints in the quality control and integrated manufacturing capabilities.
This cultural gracility, which is unfortunately even evident in many Western Europe SMEs, is explosively evident in NIS Countries for the lack of subcontracting between great industrial organisations and SMEs.
So the so called "fall-out" and "breaking-off" from great enterprises should be in some way helped by the creation of Innovation Technology Transfer Centres (ITTTC), which should act as promoters and trainers of innovation technology.
These Centres should be capable of generating an easy and friendly interaction with the correspondent Centres in EC; from here the expression of the creation of a "common european technological language", which, based on the use of common or similar transversal technologies, such as CAD, CAM, CIM and information technology in general, will allow an easy and efficient networking and the creation of a common european technological behaviour (that is a kind of concertated action in technological approach without imposing any standardization).
The innovation technology transfer by ITTTC will be also a way of commercialising the results of Russian science and technology and a tool to protect the interests of Russian scientists and engineers when they are engaged in legal agreements, for instance by establishing standards of international contracts of agreement.
This will be also a good way of activating subcontracting, that is one of the best form of cooperation, because the economic risk for a foreign partner can be reduced by operating in a common technological structure, so that Russia can offer subcontracting services at competitive prices.
Moreover, the above action will ensure the employment of NIS's scientific and technological capabilities in the NIS territory, allowing their transfer in 'real time' to SMEs; in fact only the synergistic link among both Western and Eastern (NIS) innovation technology capabilities and the NIS SMEs will support the development of the small and medium business in the NIS, that means its economic growth. In fact, it

is generally felt that the efficiency of R&D cooperation increases when it is developed in the framework of plans to support manufacturing production in Russia.
Standardisation and quality control are also important issues to be supported in NIS, especially in Russia; it should be pointed that the former Soviet Union had developed an infrastructure for standardization and normalization technology, but the lack of resources prevents making good use of it and this infrastructure is deteriorating quickly. Standardization, certification and quality control, which are all important aspects of industrial infrastructures, can be used for supporting the integration of Russia within the world market, especially an integreted European Technology Centres Network will be activated. This will be a further area of the ITTTC, which can be used in helping the process of standardization, improvement of product quality and warranty.

4. INTEGRATED SYSTEMS OF INNOVATION CENTRES:

4.1 General Considerations.

The ITTTC could coordinate the involvement of both Eastern and Western Innovation Centres grouped in "clusters" by fields of interest, that will ensure an operative link for the diffusion of innovation by scientific cooperation and supporting the local industry and public administration in different sectors.
The ITTTC could also carry out an important role to bridge the technological gap among the Western and Eastern Countries supporting the linking for a cooperation among Centres active especially in NIS and EC Science Organizations in order to transfer innovation technology to the SMEs now developing in the NIS territory.
In fact, the NIS SMEs need to improve quality and level of added value by increasing the innovative aspects, especially regarding design and manufacturing processes, that is the key of the future competitiveness on European level; the SMEs of NIS, such as those of EC, will strongly base their success on the capability of using advanced design technologies (CAD/CAE) and advanced manufacturing processes (CAM/CIM).
The urgency and necessity to proceed according to the above actions are further increased by the fact that almost all the great enterprises in West Europe are using these advanced technologies; hence, there is the risk that the SMEs in NIS territory will be cut out from the subcontracting activities because of their lack of knowledge in these techniques.
So, it is extremely urgent and strategically important to introduce the culture and know-how of computer design and computerised advanced manufacturing processes in the SMEs of the NIS, moreover it is quite evident that this knowledge should be developed on the base of a "Common European Technological Culture".
In fact, the high level of research capabilities in NIS Countries needs a well organised networking as a synergistic tool for a real cooperation with Western European Research Centres.
Therefore, the main task it is, again, to create a Common European Technological Language, especially in the transversal information technology, that will allow to diffuse and widespread the possibility of synergistic contracting within the European

area avoiding that Eastern parts of Europe, especially the most periferic ones, will be delated in activating efficient processes of information exchange development and manufacturing cooperation.
A widespread use and creation of a Common European Technological Language, could create an enormous advantage for European Community, including East side, because, thanks to this "Common European Language", the EC could become the greatest system user of the world in advanced design manufacturing technologies, with the possibility of defining the technical specifications of future systems in the area of CAD/CAM and CIM on the worldwide market.

4.2 Specific Items and objects of ITTTC.

The main objects of the ITTTC can be identified in the following:

1) to develop a kind of Common European Technological Language in the field of technologies (CAD/CAE/CAM/CIM) by creating a Network of ITTT Centres strongly and efficiently linked, thanks to a telematic network;
2) to constitute a technological reference in the field of CAD/CAE/CAM/CIM for the NIS SMEs by diffusing the designing and manufacturing computer tools used by great European organizations; this will allow subcontracting to NIS SMEs with the best cost/efficiency ratio;
3) to act as synergistic link between different technological levels (the culture generators, Universities and big enterprises, and the beneficiaries, NIS SMEs) and similar technologies which, although different in their specializations (e.g. electronic and mechanical CAD/CAM/CIM), could offer impressive advantages in being linked, at least to exchange data bases, within a Common European Network;
4) to increase control over international standards of the quality certification;
5) to establish this European technological knowledge as the strongest system user on international market to become leader in specifying performances and characteristics of CAD/CAE/CAM/CIM systems.

The industrial relevance of this action towards the NIS SMEs is enormous; in fact the risks foreseen for the NIS industrial structure in the next future will be:

a) the NIS SMEs, especially the most gracile and less cultured, might be completely excluded from high-tech knowledge in strategic manufacturing tools such as CAD/CAE/CAM/CIM;
b) most European SMEs are very active in operating with the above trasversal technologies, and the NIS SMEs will be unable to compete with them in subcontracting from great enterprises, because of their lack of well equipped infrastructures for advanced design techniques executed by CAD technologies (in electronic as well as mechanical sectors and many others like textiles, clothing, forniture, building);
c) the NIS SMEs will continue to organize their work and their manufacturing control on naive methodologies or by just taking from government R&D laboratories their "product know-how" without a capability of being integrated

with great enterprises advanced processes (CAM-CAE-FMS-CIM);

d) a substantial waste of resources in capital investments for acquiring Hw and Sw tools (mainly manufactured in USA and Japan) could happen with great dispersion and a high fragmentation of know-how (especially among Eastern Countries: Russia, Ukraine, Belorus, Moldavia etc.).

In case of activation of the ITTTC there will be various advantages which can be resumed in:

a) low cost and high efficiency diffusion within SMEs of know-how in advanced computer controlled technologies, in designing (CAD/CAE) as well as in manufacturing control (CAM/CIM);
b) great improvements in efficiency for subcontracting from great enterprises to NIS SMEs with a strong action of driving and pushing NIS SMEs towards advanced tools for improving quality and added value of products;
c) soft, but solid, integration of already existing CAD/CAE/CAM/CIM European Centres which, exchanging via an efficient telematic link "methodologies and behaviour", will create a "European Technological Language" in the area of the above technologies;
d) creation, thanks to the link between computer design (CAD/CAE) and manufacturing (CAM/CIM), of a European concertated technological know-how, which will constitute the greatest System User on world basis for these strategic technologies, with all the consequent advantages in power and credibility for specifying the performance criteria and characteristics of future tools in these fields.

Besides, it has to be underlined that this action foresees an open structure that could involve the majority of European SMEs and CAD/CAM/CIM Centres, already existing or in start phase, so that the number of SMEs to be involved in this action could be expected to growth in to many thousands.

5. CONCLUSIONS

An important action to support the economic growth of the former Soviet Union is the development of an integrated system of innovation centres operating in transversal technologies (information technologies, CAD/CAM/CIM etc.) defined by the acronym ITTTC (Innovation Technology Transfer and Training Centres);
this will allow:

1) grow-up of technological culture by continuing training, based on real service (projects, design, etc.) and the teaching of "learning to learn";
2) individuation of common tasks based on the use of new strategic, horizontal technologies (CAD/CAM, information technology and telematic);
3) quality certification by advanced technological behaviour;
4) creation of specific "niches" areas for NIS and EC SMEs within CEC R&D (e.g. Sprint, Esprit, Craft, Brite) and Eureka Programmes, to improve the interaction with great enterprises.

Last but not least, the above action will promote a new cultural attitude which, awarding the cooperation among European structures, will enhance the capability of creating new enterprises, supplying a fondamental task of job creation, derived by a solid and deep know-how in advanced technology.

INNOVATION ACTIVITY AND FIRM SIZE

BENIAMINO QUINTIERI
Faculty of Economy and Business Administration
Rome University "Tor Vergata"
Rome, Italy

The problem of measuring the level of innovation is a complex one, especially in small e medium enterprises (SMEs). In fact, the data does not allow one to measure change directly the innovation activity. The solution adopted thus far involves estimating innovation activities via statistical measurement of its determining factors and of its supposed outputs. R&D expenditures have been by far the most used proxy for technological innovation R&D is considered as an indicator of inputs, implying that R&D is the main activity which leads to innovation. The use of such indicator has been criticized because R&D may constitute a reliable proxy for innovation only for large-sized firms and fail to reflect a whole series of technological activities prevalent in medium and small firms. Patents have been also employed as an alternative indicator of the innovative activity. Although generally regarded as less biased toward larger firms, patents are not a measure of innovations but rather a measure of invention. For these reasons other measure have been considered like the number of innovations or innovation costs. Because data by size-class are usually limited these measure are obtained from survey data or ad-hoc studies. Data on R&D suggest that the number of SMEs performing R&D relative to the total number of SMEs is lower than the equivalent ratio in large enterprises. For example in French industry less than 5% of SMEs perform formal internal R&D against 50% of largest enterprises. As it emerges from Table 1 and Figure 1, R&D activities are generally concentrated in largest enterprises. Also data on patents show that the innovation activities increase with the size of the firms.

	A	B	C	D
Belgium	-	33	36	69
Denmark	86	39	41	76
France	77	18	21	67
Germany	-	12	17	62
Ireland	-	82	-	83
Italy	76	19	23	81
Netherlands	-	16	13	72
Portugal	65	37	-	80
Spain	-	36	42	83
United Kingdom	-	9	6	65

C. Corsi (ed.), Science and Innovation as Strategic Tools for Industrial and Economic Growth, 15-19.

Source: The European Observatory for SMEs, 1994

Tab. 1 Share of SMEs in R&D activities compared to their share in employment (1990)

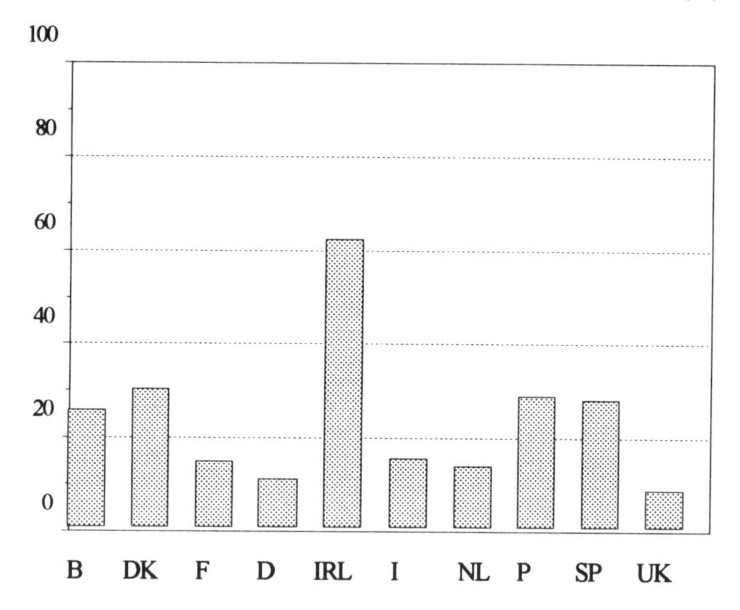

Fig. 1 Ratio between SMEs' R&D share and share of SMEs in employment

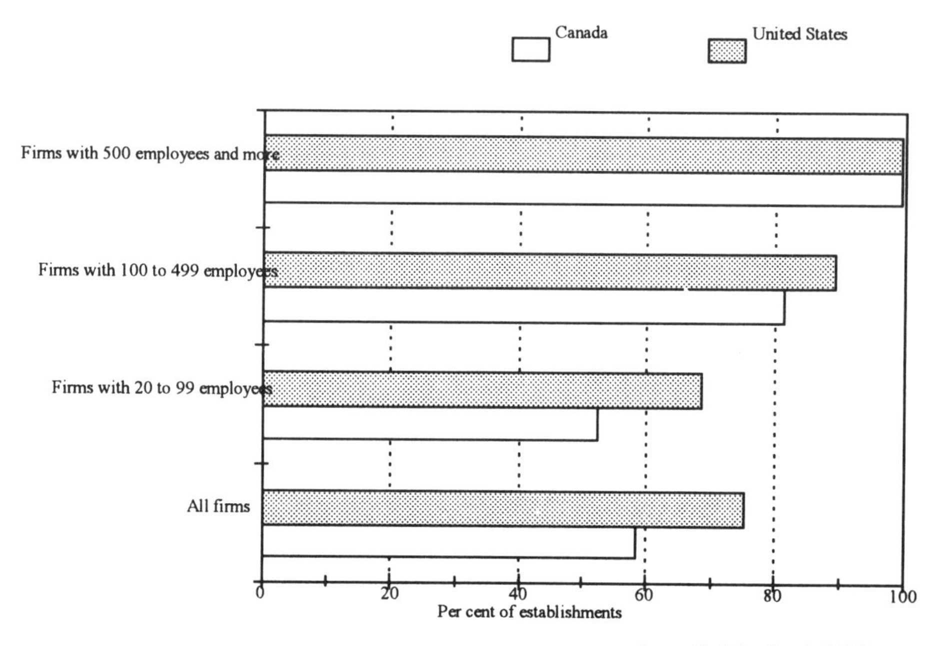

Source: Statistics Canada (1991).

Fig. 2 Use of computerised production technologies in manufacturing firms in Canada and the United States, 1989

Comprehensive surveys on innovation within enterprises are quite rare. Available evidence (Table 2) seems to indicate that the trend to innovate tends to increase with size of the enterprise. As far as new production technologies are concerned, SMEs lag further behind large firms but the lag is tending to become less. Various reasons have been put forward to explain the SME lag and the variable penetration of technologies according to sector and firm.

Netherlands	France		Germany	Italy	Luxembourg			
Size class		Pd1	Pr2		Pd	Pr	Pd	Pr
0-9	n.a.	n.a.		n.a.	40	33	42	28
10-19	n.a.	n.a.		n.a.	49	50	50	58
20-49		32	35	63		64	77	
50-99				75		67	74	
100-199	55	53	51	82	89.5	79	73	79
200-499	79	63	59	83.5				
Total LSEs	93	78	75	89	100	100		
Total	60	66	63	69	50	46	51	47

Tab. 2 Number of enterprises being innovative as a percentage of the total number of enterprises by size class

A recent study of OECD3 has shown that behind innovation or the introduction of new production technologies is scientific and technological information, the essential resource which has become as important and as strategic than capital. A recent survey in the Netherlands (Table 3) showed that the difficulty of finding technological information was one of the three parameters, together with lack of capital and costs of monitoring, in respect of which then was a significant difference between the small and the larger firms in terms of innovation.

Various attempts to construct more reliable indicators of innovative activity have been done employing mainly data from survey data at country level. In Italy a new indicator has been used4 and it is based on total costs sustained by the firms for innovative activity as a whole. The main advantage of this indicator is the fact that it reflects the efforts made by the firms in carrying out a particularly broad spectrum of innovative activities. The inclusion of new innovative sources produces then the effect of reducing the bias associated with the use of the more traditional technological indicators. Table 4 presents data for a survey of Italian firms, broken down by firm size regarding the

---

[11] 1Pd: Product Innovation

2Pr: Process Innovation

[3] OECD, "Small and medium-sized enterprises technology and competitiveness", Paris, 1993.

4 See Archibugi-Evangelista-Simonetti, "Concentration, firm size, and innovation, evidence from innovation costs", in *Rivista Internazionale di Scienze Sociali*, n. 3, 1993.

division of innovation costs into four categories: R&D, Design and Engineering, Production Investment, Marketing. R&D expenditures accounts for 21.4% for the largest size group and barely 7.4% for the smallest, providing confirmation of how much this indicator underestimates the innovative efforts of the smaller firms. Conversely an important role is played by Production Investment with its relative weight displaying an inverse relation with firm size: 73% for the smallest size group and 43.5% for the largest one.

	Size of firms by number of employees					
Problems	10-19 n=75	20-49 n=294	50-99 n=330	100-199 n=264	200-499 n=176	>500 n=112
Lack of capital	58.37	47.3	38.8	33.7	29.0	26.8
Difficulties in predicting demand	57.3	47.6	49.7	59.1	48.3	50.9
Apparent costs of developing innovation	37.3	36.1	32.7	33.3	35.8	30.4
Problem of adapting the marketing function	25.3	25.5	27.6	26.5	25.0	25.9
Costs of monitoring future applications	29.3	26.9	27.9	20.5	19.3	11.6
Difficulties in finding technological information	24.0	20.4	26.1	18.2	22.2	8.9
Employee skills	24.0	20.7	21.2	20.1	12.5	12.5
Government standards	13.3	8.8	11.2	10.6	13.6	13.4

Source: OECD, "Small and medium-sized enterprises: technology and competitiveness", 1993.

Tab. 3 Relative significance of the different problems encountered by firms in innovation (Netherlands) Percentage

Firm Size (no. of employees)	R&D	Design and Engineering	Production Investment	Marketing	Total
20-49	7.4	15.1	73.1	4.4	100.0
50-99	9.6	17.5	68.2	4.7	100.0
100-199	10.6	18.3	66.2	4.9	100.0
200-499	14.3	16.1	64.2	5.4	100.0
500 and over	21.4	29.5	43.5	5.6	100.0
Total	17.9	25.2	51.5	5.4	100.0

Source: CNR ISTAT. Sample of 8220 firms

Tab. 4 Breakdown of innovation costs by firm size

Table 5 reports evidence on innovation costs broken down at industry level. The sectors that show a larger expenditure on innovation are those generally associated with higher technological opportunities. The six sectors to report the highest innovative intensity are Office machinery and Computers, Transports and aircraft, Rubber and Plastics, Motor-vehicles, Precision instruments, Electronics. All together these sectors cover 55% of the total innovation costs and 67% of the R&D activities.

The results concerning the relationship between size and innovative activity confirm the existence of a positive relationship between these two variables both at an aggregate level and for the group of the highly innovative sectors. At the same time however results indicates that in the traditional industries such as Metal products, Food, Textile, Leather, Footwear, firms of small or medium size spend on innovation more than large firms, even when large firms spend proportionally more on R&D.

These results lead to the conclusion that in determining the contribution of SMEs to innovative activity a crucial role is played by sectoral differences in terms of technological regimes.

	No. Of firms	Innovation costs (%)	R&D exp. (%)	Innovative intensity (innovation costs per employee) R&D expendit.	Design and Engineering	Produc. invest.	Mark	Total innov. costs
Petrochemicals	17	0.91	0.75	2.6	2.3	13.1	0.3	18.3
Metals	137	4.83	1.85	0.8	5.4	5.9	0.5	12.6
Non metals, Minerals	495	4.94	1.38	0.9	1.7	16.4	0.5	19.6
Chemical	409	7.74	13.49	4.5	2.4	6.9	1.3	15.1
Synthetic fibres	9	0.46	0.31	1.3	1.1	8.5	0.2	11.1
Metal products	864	3.37	1.29	0.7	1.7	7.6	0.4	10.4
Mechanical Machinery	1166	9.37	8.78	2.2	4.6	6.4	0.6	13.9
Office machin. computers	11	10.11	20.41	25.0	19.3	17.8	10.0	72.0
Electrical, Electronics	526	13.99	19.69	5.1	7.1	8.2	0.8	21.2
Motorvehicles	155	16.09	14.31	4.2	6.5	16.3	0.5	27.5
Other transport	88	7.69	1.74	1.1	13.7	13.9	0.2	29.0
Precision instruments	110	1.37	1.65	5.3	7.0	11.3	2.1	25.7
Food	268	2.38	0.50	0.5	0.8	11.7	0.6	13.7
Sugar, Drinks	177	2.10	1.34	1.0	1.2	6.6	0.3	10.3
Textiles	610	2.99	1.11	0.7	0.7	8.6	0.3	10.3
Leather	97	0.20	0.05	0.3	0.8	6.4	0.3	7.7
Footwear, Clothing	343	0.64	0.20	0.2	0.6	2.4	0.2	3.5
Wood, Furniture	469	1.11	0.50	0.8	1.2	7.5	0.6	10.1
Paper, Printing	369	3.86	0.99	0.6	0.6	12.6	0.7	14.5
Rubber, Plastic	389	5.51	9.16	8.1	7.0	10.0	3.1	28.2
Other manufacturing	130	0.32	0.24	1.3	1.6	6.9	0.4	10.2
Total	6,839	100.00	100.00	3.2	4.6	9.6	0.9	18.3

Source: *Indagine CNR-ISTAT*.

Tab. 5 Innovation cost by industry (sectoral average values)

MANAGING TECHNOLOGY IN LARGE MULTIBUSINESS FIRMS

VEREDICE G.,
R&D V. Director, Finmeccanica
Rome, Italy

1. INTRODUCTION TO THE PRESENTATION

The subject of this discussion is the strategic use of technology in a multibusinees firm that supplies worldwide a variety of complex assembled products to customers in industry and government. To make comprehensive strategic decision, the senior management of such an organization needs an intimate understanding of the evolution of the technology base underlying its products. In a situation in which technologies are not stable and predictable, the development of the firm's technology base often leads business strategy, rather than merely support it.

Technological development may be pursued with different means: in house R&D, interfirm cooperation, and cooperative R&D efforts with universities and government agiencies. In the first section of my presentation, I describe the process of technology planning used at the corporate headqarters of Finmeccanica to coodinate these means and to monitor the R&D activities carried out at the various divisions and subsidiaries, accordin to the fully decentralized approach to R&D adopted in our organization. The planning process is focused on the technologies that are embodied in the performance of our products and in the processess used to design, manufacture, and service our products. On a regular basis, we analyse some of our business units to assess their comptetive position in those technologies that are the key to stay successfully in the industry, or that are emerging as capable to better satisfy the customers needs. We also address specific technologies to evaluate their potential across different business units, and sometimes provide seed funding to encourage basic R&D efforts at the "centre of excellence" we identified in some areas of our organization. Finally, we keep track of each business unit's R&D projects, both current and planned, to assess whether they are consistent with the strategic goals of the business unit.

In the second section of the presentation, I address the increasing use of interfirm cooperation as a means to pursue technology development. Although sharp distinctions are not easy to make, I try to categorize industrial alliances according to the three dimensions of scope, orientation, and type of legal arrangement. Alliances may pursue strategic goals, i.e. be concerned with the firm's long-term competitive position on product market, or they may focus on short-term efficiency objectives in specific segments of business operations. Alliences may be directed primarily to technology transfer and development, or they may be concerned with manufacuring and marketing.

C. Corsi (ed.), Science and Innovation as Strategic Tools for Industrial and Economic Growth, 21-34.

Finally, alliances may assume a variety of contractual modes, ranging from unilateral technology flows such as licensing to joint ventures. A number of different motives may lead to strategic technology partering, i.e., to alliances that are both strategic and technology-oriented: complementary technologies, high risks of R&D, the innovation time span, and others. I argue that in many high-tech industries, strategic alliances are predominantly technology-oriented. This conclusion is supported by empirical studies, and I illustrate some results adapted from the analyses conducted by Hagedoorn (1993) and by Hagedoorn and Schakearaad (1994) on a wide data base of alliances made over the last decade.
My final remarks concern one special case of strategic technology partnering, namely the alliance between small firms and large firms. It is my opinion that these alliances may be more successful than others, since they are more ofthe based on specific and complementary assets provided by each partner.

CONTENTS

- TECHNOLOGY PLANNING
- ALLIANCES

TECHNOLOGY PLANNING
TECHNOLOGICAL DEVELOPMENT

- THE CAPABILITY TO MASTER COMPLEX TECHNOLOGIES IS TODAY A KEY FACTOR TO COMPETE SUCCESSFULLY IN MANY INDUSTRIES.

TECHNOLOGY PLANNING
TECHNOLOGICAL DEVELOPMENT: TOOLS

TECHNOLOGICAL DEVELOPMENT MAY BE PURSUED THROUGH:
- IN HOUSE R & D
- INTERFIRM ALLIANCES
- COOPERATION WITH UNIVERSITIES AND GOVERNMENT AGENCIES

THE SCOPE OF TECHNOLOGY PLANNING

THE COORDINATION OF THESE TOOLS IS THE SCOPE OF TECHNOLOGY PLANNING.

THE PROCESS OF TECHNOLOGY PLANNING

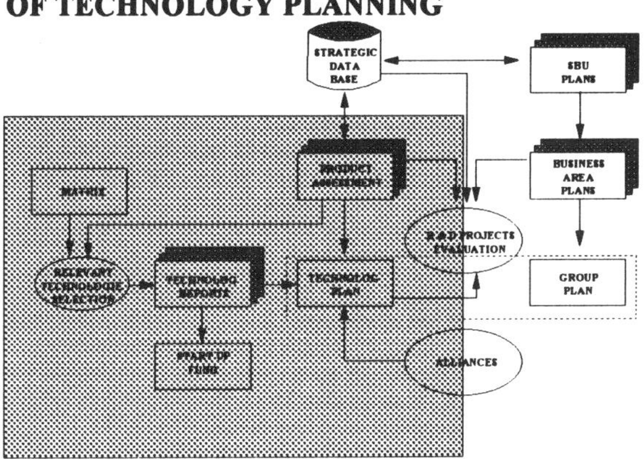

TECHNOLOGY PLANNING

DEFINING TECHNOLOGIES

ELEMENTARY TECHNOLOGIES IDENTIFY THE CAPABILITIES WHICH UNDERLIE THE DESIGN, FABRICATION, INSTALLATION AND SERVICE OF PRODUCTS. THEY HELP DEFINE A FIRM'S OR COMPETITIVE POSITION.

TECHNOLOGY PLANNING

CLASSIFYING TECHNOLOGIES

- WE GROUPED ELEMENTARY TECHNOLOGIES ACCORDING TO A TREE STRUCTURE BASED ON PROXIMITY OF CONTENTS.
- WE ESTABLISHED ELEVEN TECHNOLOGY AREAS AS BOTTOM BRANCHES. ABOUT ONE HUNDRED "REFERENCE" ENTITIES FOLLOW NEXT UP THE TREE.

TECHNOLOGY PLANNING
CLASSIFYING TECHNOLOGIES, AN EXAMPLE

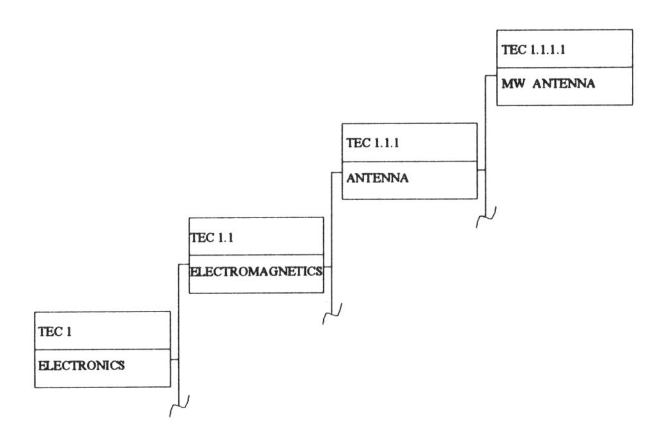

TECHNOLOGY PLANNING
TECHNOLOGY MATRIX

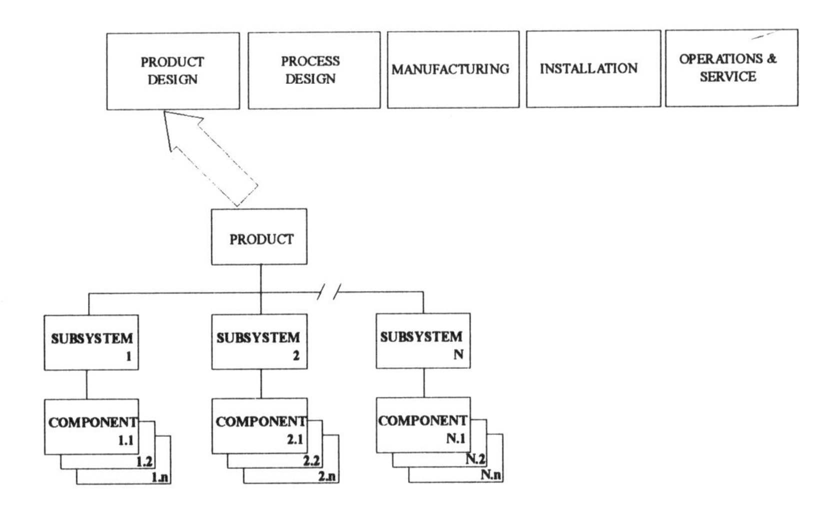

TECHNOLOGY PLANNING
TECHNOLOGY MATRIX, AN EXAMPLE

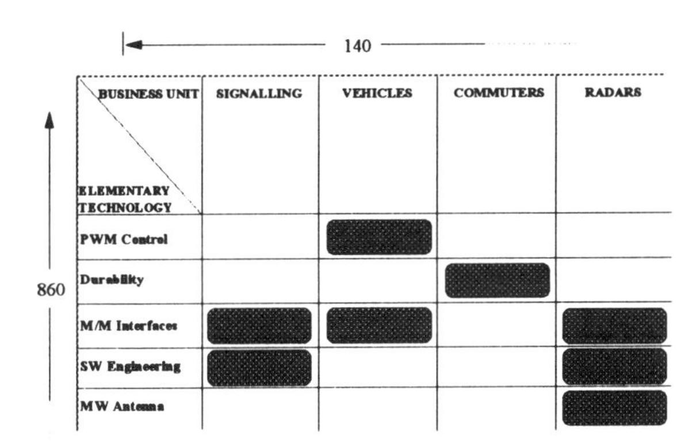

TECHNOLOGY PLANNING
RELEVANT TECHNOLOGIES

WE ASSESSED THE COMPETITIVE IMPACT AND MATURITY LEVEL OF THE ELEMENTARY AND REFERENCE TECHNOLOGIES TO IDENTIFY THOSE CRITICAL TO EACH BUSINESS UNIT.

BY USING MULTIBUSINESS IMPACT AND ACCUMULATED INTERNAL COMPETENCES AS ADDITIONAL RANKING CRITERIA, WE OBTAINED A SET OF CRITICAL TECHNOLOGIES AT CORPORATE LEVEL.

TECHNOLOGY PLANNING
RELEVANT TECHNOLOGIES

- MULTIBUSINESS IMPACT
- COMPETITIVE IMPACT
- MATURITY LEVEL
- INTERNAL COMPETENCES

TECHNOLOGY PLANNING
RELEVANT TECHNOLOGIES, QUESTIONNAIRE

FIRM [ ] BUSINESS [ ] MAIN COMPETITORS ....
DATE [ ] PRODUCT/S [ ] ....
....
....

Reference Technology	Elementary Technology	Technology Expertise	Technology Content	Trend	Impact Potential (rank the key factors)				Competitive Impact	Maturity	Technological Competitive Posit.	Person.	Investment		Common to business
					Performance ( )		Other ( ) specify						Past	Future	

Attributes
Technology expertise (A-Application Technologist, T-Technologist); Technology Content (1 Low ... 5 High); Trend (RR - Strong Reduction, R-Reduction, S-Steady, I-Increase, II-Strong increase).

Impact Potential (1 Low .... 5 High); Maturity (E-Embryonic, AG-Accelerating Growth, DG-Decelerating Growth, M-Mature)
Technological competitive position (WE-Weak, TE-Tenable, Fa-Favorable, ST-Strong, DO-Dominant). Annual average personnel (man-years)
Annual average investment (USD thousands)

TECHNOLOGY PLANNING
TECHNOLOGY REPORTS

EACH REPORT ADDRESSES ONE SPECIFIC TECHNOLOGY.
A PANEL OF INTERNAL EXPERTS IS SET UP TO:

- ASSESS INTERNAL KNOW-HOW AGAINST STATE-OF-THE-ART
- ANALYSE ONGOING R&D PROJECTS
- HIGHLIGHT POSSIBLE R&D TRENDS
- DEFINE PROJECTS TO BE LAUNCHED:
  - AT BUSINESS UNIT LEVEL
  - AT GROUP LEVEL
- IF NEEDED, AN EXTERNAL EXPERT ASSISTS PANEL WORK.

TECHNOLOGY PLANNING
PRODUCT ASSESSMENTS

ASSESSMENTS ARE FOCUSED ON SPECIFIC PRODUCTS. A PANEL INCLUDING THE PRODUCT MANAGER, A REPRESENTATIVE OF CORPORATE STAFF, AND AN EXTERNAL ADVISOR IS SET UP TO:

- IDENTIFY PRODUCT-RELATED TECHNOLOGIES

- EVALUATE THEIR IMPACT ON COMPETITIVENESS
- ASSESS COMPANY'S POSITION VS COMPETITORS'
- SUGGEST ACTIONS TO BE TAKEN

TECHNOLOGY PLANNING
START - UP FUND

- AIMED TO:
- SUPPORT START-UP OF MULTI-DIVISION R&D PROJECTS
- STRENGTHEN CENTERS OF EXCELLENCE

TECHNOLOGY PLANNING
TECHNOLOGY PLAN

THE TECHNOLOGY PLAN, THE ANNUAL OUTPUT DOCUMENT OF THE PLANNING PROCESS, PROVIDES SENIOR MANAGEMENT AT THE CORPORATE HEADQUARTERS WITH A SURVEY OF THE R&D ACTIVITIES IN THE FIRM'S DIVISIONS AND SUBSIDIARIES

TECHNOLOGY PLANNING
TECHNOLOGY PLAN, APPROACH

- IDENTIFY AND PROMOTE INTERNAL COMPETENCES
- ASSESS CURRENT AND PROJECTED R & D EFFORTS *VIS-A-VIS* MARKET GOALS

TECHNOLOGY PLANNING
TECHNOLOGY PLAN, CONTENTS: OVERALL INDICATORS

HISTORICAL, CURRENT AND PROJECTED DATA:

- R & D PERSONNEL AND EXPENDITURES
- R & D FUNDING SOURCES

DATA ARE GATHERED AT BUSINESS UNIT LEVEL

TECHNOLOGY PLANNING
TECHNOLOGY PLAN, CONTENTS: FACILITIES

HISTORICAL AND CURRENT DATA ON:
- R & D "FACILITIES" (DESIGN TEAMS, SOFTWARE DEVELOPMENT TEAMS, LABS, TEST FACILITIES, ETC.)

SPECIAL ATTENTION TO:
- CENTERS OF EXCELLENCE, BOTH ACTUAL AND POTENTIAL

TECHNOLOGY PLANNING
TECHNOLOGY PLAN, CONTENTS: COMPETITORS

HISTORICAL AND CURRENT DATA ON COMPETITORS' R & D EXPENDITURES AND REVENUE AT:

- GROUP LEVEL
- BUSINESS AREA LEVEL (AEROSPACE, DEFENCE, ENERGY, TRANSPORTATION AND AUTOMATION)

TECHNOLOGY PLANNING
TECHNOLOGY PLAN, CONTENTS: SBU ANALYSIS

EACH BUSINESS UNIT IS ANALYSED. R & D DATA ARE GATHERED AT PROJECT LEVEL TO PROVIDE:

- TECHNOLOGICAL POSITION, BOTH PRESENT AND PROJECTED
- OVERALL R & D EXPENDITURES
- (A + B) R & D AS A PERCENTAGE OF OVERALL EXPENDITURES
- R & D EXPENDITURES AS A PERCENTAGE OF REVENUE

TECHNOLOGY PLANNING
IN HOUSE R&D - CLASSIFICATION

- TYPE "A": PRELIMINARY RESEARCH
  Long term, Highly innovative, High risk
- TYPE "B": FINALIZED RESEARCH
  Mid term, Innovative, Medium risk
- TYPE "C": DEVELOPMENT
  Near term, Slightly innovative, Low risk

ALLIANCES

Firms increasingly extend the reach of their operations through a variety of cooperative agreements.

CATEGORIZING ALLIANCES, I

Alliances can be categorized by orientation and scope of motives, and by mode of legal arrangement.

CATEGORIZING ALLIANCES, II

Orientation distinguishes alliances focused on technological innovation from those directed to improving market access and positioning within market structure.

Scope is related to time perspective. Strategic alliances have a long-term interest towards market power; economizing alliances are mainly concerned with increasing efficiency in the short-term.

CATEGORIZING ALLIANCES, III

Sharp distinctions are however impossible.

The motives underlying each alliance are often of a mixed nature. Moreover, the same motive may have different orientation and scope for each partner.

CATEGORIZING ALLIANCES, IV

Some modes of cooperation are likely to be strategically motivated; others tend to be more often associated with cost-economizing motives.

Also, complex modes such as joint ventures aim at both market and technology objectives; simple contractual arrangements are more often technology-oriented.

ORIENTATION AND SCOPE OF INTERFIRM COOPERATION

MOTIVES FOR INTERFIRM COOPERATION

MODES AND SCOPE OF INTERFIRM COOPERATION

Modes	Description	Scope
Joint venture/research corporation	Combination of interests in a "distinct" firm	Often strategically motivated,
Joint R&D agreement	Joint undertaking of R&D projects with shared resources	i.e. aimed at long-term perspective of market position of companies
Equity investment	Purchase of equity stakes (minority and cross holdings)	involved
Technology exchange agreement	Technology sharing, cross-licensing, and mutual second-sourcing	More oriented towards cost
Customer-supplier relationship	Contract regulated collaboration in either production or research	economizing, i.e. control of transaction or operating costs of
Unilateral technology flow	Second sourcing and licensing	companies involved

Derived from Hagedoorn and Schakenraad, Strategic Management Journal, Vol. 15, 291-309 (1994)

MODES AND ORIENTATION OF INTERFIRM COOPERATION

	Frequency (*) of Technology-Oriented Motives	Frequency (*) of Market-Oriented Motives
Complex Modes	1	1.4
Contractual Arrangements	3.1	0.8

(*) Frequency relative to technology-oriented motives in complex modes
Adapted from Hagedoorn, Strategic Management Journal, Vol. 14, 371-385 (1993)

ALLIANCES IN HIGH-TECH INDUSTRIES

THE TECHNOLOGY INTENSITY OF AN INDUSTRY IS REFLECTED IN THE ORIENTATION AND SCOPE OF THE ALLIANCES MADE WITHIN THAT INDUSTRY.

STRATEGIC TECHNOLOGY PARTNERING IN HIGH-TECH INDUSTRIES

STRATEGIC TECHNOLOGY PARTNERING IS AIMED AT JOINT INNOVATIVE EFFORTS THAT CAN HAVE A LASTING EFFECT ON THE PRODUCT-MARKET POSITIONING OF PARTICIPATING COMPANIES.

IN MANY HIGH-TECH INDUSTRIES, TECHNOLOGY PARTNERING IS THE PREVAILING FORM OF STRATEGIC ALLIANCE.

MOTIVES FOR STRATEGIC ALLIANCES IN HIGH-TECH INDUSTRIES

	Monitor Technology/Market entry	Complement Insufficient Financial Resources	Reduce Innovation Time Span	Exploit Jointly Complementary Technologies	Share High Cost/Risks of R&D	Influence Market Structure	Perform Together Basic Research
Biotechnology	●●	●●	●●●	●●●	●	●●	●●
New Materials	●●	●	●●●	●●●	●	●●●	●●
Computers	●●	●	●●●	●●●	●	●●●●	●
Industrial Automation	●	●	●●●	●●●●	●	●●●	●
Microelectronics	●	●	●●●	●●●	●	●●●●	●
Software	●●	●	●●●	●●●	●	●●●	●
Telecommunications	●●	●	●●●	●●●	●●	●●●	●
Aerospace/Defense	●	●	●●●	●●●	●●●	●●	●
Heavy Electric/Power	●●	●	●●	●●●	●●●	●●●	●
Instruments	●●	●	●●●	●●●	●	●●●	●

● never or seldom
●● sometimes
●●● often
●●●● very often

Adapted from Hagedoorn, Strategic Management Journal, Vol. 14, 371-385 (1993)
Results refer to 3310 alliances made during the period 1980-1989

TECHNOLOGY-ORIENTED ALLIANCES PREVAIL IN HIGH-TECH INDUSTRIES

	Technology/Market Ratio
Biotechnology	TT
New Materials	T
Computers	M
Industrial Automation	T
Microelectronics	M
Software	T
Telecommunications	M
Aerospace/Defense	TT
Heavy Electric/Power	T
Instruments	T

TT	Technology driven
T	Mixed motives, but technology prevails
M	Mixed motives, but market prevails
MM	Market driven

Adapted from Hagedoorn, Strategic Management Journal, Vol. 14, 371-385 (1993)
Results refer to 3310 alliances made during the period 1980-1989

DOES FIRM SIZE MATTER TO A SUCCESSFUL ALLIANCE?

LARGE FIRMS SEEM TO HAVE BETTER AND MORE OPPORTUNITIES TO SEEK EXTERNAL LINKAGES.

HOWEVER, SIZE IS SELDOM A VIRTUE IN CREATIVE BUSINESSES. STRATEGIC TECHNOLOGY PARTNERING WITH SMALLER PLAYERS IS A MEANS FOR LARGE FIRMS TO TAKE ADVANTAGE OF THE LATEST INNOVATIONS. THIS FORM OF COOPERATION BETWEEN SMALL AND LARGE FIRMS IS A COMMON PATTERN IN MANY HIGH-TECH INDUSTRIES.

BEYOND TRADITIONAL RELATIONSHIPS, I

IN HIGH-TECH INDUSTRIES, THE TRADITIONAL ADVERSARIAL MODEL OF RELATIONS BETWEEN SMALL SUPPLIERS AND THEIR LARGE CUSTOMERS DOESN'T HOLD.

LARGE FIRMS JOINTLY DESIGN NEW COMPONENTS WITH THEIR SUPPLIERS, ENTER INTO JOINT VENTURES TO DEVELOP A NEW TECHNOLOGY, OR SPIN-OFF TECHNOLOGIES DIFFICULT TO COMMERCIALIZE IN-HOUSE.

BEYOND TRADITIONAL RELATIONSHIPS, II

AS A RESULT, LARGE FIRMS INCREASE THEIR SPEED AND FLEXIBILITY, WHILE SMALL SUPPLIERS WITH VALUABLE TECHNOLOGICAL EXPERTISE ARE GIVEN MUCH MORE ROOM TO MANEUVER THAN SUPPLIERS TYPICALLY HAVE.

WHY SHOULD LARGE AND SMALL FIRMS COOPERATE?

LARGE FIRMS

- CAPABLE TO MANAGE TRANSITION FROM DRAWING-BOARD TO MARKET
- CAPABLE TO ASSESS AND COMBINE DIVERSE KNOWLEDGE
- COMPLEMENTARY ASSETS SUCH AS DISTRIBUTION CHANNELS, SERVICE NETWORK, BRAND IMAGE, ETC...

SMALL FIRMS

- SPECIALIZED KNOWLEDGE
- CREATIVE APPROACH TO MARKET OPPORTUNITIES
- FLEXIBLE, FAST, AND RISK-TAKING

CULTURES OF TECHNOLOGICAL INNOVATION IN RUSSIA

YAKOV M RABKIN
President of Periscence Inc.,
Montreal Canada

1. INTRODUCTION

Technological innovation is known to depend on diverse factors: economic development, scientific prowess, political environment, social mobility among employment sectors, etc. Other factors are less tangible.

This paper considers one of such factors, namely cultural values related to the process of technological innovation. While cultural values cannot be entirely independent variables, they exercise a potent influence on the process of innovation, albeit in a less univocal or direct manner.

This is why it is important to look at cultural values of technological innovation in Russia, a country which underwent two momentous revolutions in its political and economic structures, in the late 1910s (consolidated in 1929, the Year of the Great Break) and in the early 1990s.

In the course of both of these changes political and economic leaders paid consistent lip service to the importance of innovation.

However, the very intensity and frequency of their admonitions betrayed a deep-seated frustration with what they deemed was a slow pace of reforms and innovation. They blamed the passivity of the population, its uneducated attitudes rooted in history. This paper will attempt not only to verify the validity of these complaints, but to cast a broader view on what can be called "cultures of technological innovation", i.e. a complex of modes of behaviour and customs, as well as values underlying them.

Modern science has assumed in Western contexts a variety of functions, including, perhaps, the most important role of a catalyst of technological change valued by a broad class of economic players. This role was largely absent in Russia and, in spite of appropriate ideological verbiage, in the Soviet Union. The number of economic players was limited under the empire and was reduced to one - the State - under the communists. Technological innovation rarely stemmed from domestic scientific efforts, and there emerged the impression that the mighty infrastructure of science drained rather contributed to the country's material resources. Throughout the country's history, science did not become an economic factor comparable to its role in Western

C. Corsi (ed.), Science and Innovation as Strategic Tools for Industrial and Economic Growth, 35-50.

industrialized societies.

2. TECHNOLOGY TRANSFERS IN RUSSIAN CULTURES

Technology transfer is a concept germane to any discussion of technological development. It ought to be defined at the outset.

Technology transfer is any process that brings home foreign innovations with the purpose of catching up with other companies, industries, or countries. The effectiveness of technology transfer is contingent on the recipient's ability to absorb, diffuse and assimilate new technologies.

An analogy can be made with organ transplant: success mainly depends on the ability of the body to assimilate the new organ, not on the latter's intrinsic qualities.

The usual pattern of technology transfer to Russia has been that of importing people. The tsars had intuitively understood the principle enunciated by the late historian of science Derek De Solla Price in the 1960s that the most efficient way to bridge the chasm between science and production (as well as between different levels of technological development) is to bring about the physical movement of qualified personnel between donors and recipients.

Among the earliest known agents of technology transfer were the Italian architects assigned to the construction of the Moscow Kremlin which has since acquired the character of a quintessential example of Russian architecture. The pugnacious reformer Peter the Great was greatly influenced by foreigners in Moscow's German Colony (Nemetskaya sloboda), expanded under the rule of his less known father Alexis who has come to be credited with a more significant role in the Westernization of Russia than previously admitted.

A massive technology transfer took place in the last five decades of the Russian monarchy. Count Witte was instrumental in opening up Russia's industries to foreign investment and foreign know-how. There resulted a series of improvements in Russia's technological performance, making Russia one of the fastest growing economies on the eve of the First World War.

The October Revolution initially signaled a departure from the practice of technology transfer. Lenin's government expropriated all private enterprise, foreign or locally owned, unilaterally cancelled foreign debts, and authorized only a limited number of concessions by foreign companies willing to introduce new technologies. Such concessions amounted to less than one percent of the nationalized industrial output.

However, a decade later Stalin's five-year plans overtly admitted the technological backwardness of the Soviet Union. He consequently embarked on an ambitious programme of importing latest technological equipment which amounted to over one half of the country's total imports. These imports were made possible by the exceptionally brutal expropriation of agricultural produce resulting in millions of human lives. In spite of this cruel toll thousands of Western technologists, prompted by the crisis affecting most of the industrial world, flocked into Stalin's Soviet Union to install new equipment, train local manpower and serve as advisors to industrial managers. At the same time, resources were generously allocated to the country's expanding R&D system. The 17th Party Congress proclained in 1934 that the USSR was on its way to become the world's most advanced industrial state no longer in need of foreign technology imports.

From the 1930s on, the Soviets perceived, although seldom publicly admitted, inferiority of the Soviet technological performance.

On the face of it, the decade following the end of World War Two was a period of unbridled chauvinism in Soviet science. Histories were speedily rewritten in order to "prove" that Russians were behind all important scientific breakthroughs in the world. One does not have to be an adept of psychohistory to realize that it was the Soviets' insecurity about their scientific potential. It was further sharpened by the then haunting American monopoly on nuclear arms, that was behind the unprecedented campaign to "Russify" the world history of science. Several spy scandals involving the Soviet Union in the 1940s and 1950s also highlighted the acuity of the Soviets' interest in American science.

When more information about Soviet science reached the West after Stalin's death it became clear that the gap between the levels of Soviet and of American scientific efforts in the preceding decades had been deep indeed .

Thus it was no paradox that while the Soviets were whipping up the chauvinistic glorification of their national science in the 1940s and 50s, they continued to supply their libraries with an impressive array of American scientific journals. Even in the midst of an officially sponsored "struggle against cosmopolitism" many Soviet scientists would read English and make ample use of American scientific literature. At the same time, tight controls were maintained on foreign travel of scientists and engineers. Access to Western science and technology was of great interest to Soviet leaders. But the importance of such access was secondary to the prospect of a loss or even a diminution of its controls over Soviet population. Long-term concerns about Soviet technological performance would not counterweigh their constant apprehension about challenges to the power of Soviet leaders.

3. FUNDAMENTALIZATION AND ITS CONSEQUENCES

Soviet rulers used to believe that investment in basic research would trigger technological innovation. Science was deemed to be the motive force of development. However, since the mid-1970 research came to be perceived as a useless and expensive activity with little benefit to industry or agriculture. This was hardly the fault of science; rather it was the Soviet economic system that was largely impermeable to technological innovation, particularly from indigenous endeavours.

Fundamental science as a source of military and civilian technologies had begun to lose its significance for the government since the mid-1970s when research funding had reached a peak. It became tacitly, and since the mid-1980s overtly, acknowledged that bottlenecks in Soviet industry and agriculture were not of scientific nature, and could not be remedied by science alone. It was beleived that new social modes of production, not new scientific knowledge, were needed to awaken the Soviet economy from stagnation. The most insightful among the nomenklatura understood that the Soviet system could no longer ensure the future of their privilege. They thus began to assimilate Western approaches and attitudes with the express purpose of perfecting rather than abolishing the old system. Thus began perestroika, Gorbachev's desperate attempt to reform and improve the Communist system.

While innovation was deemed an incontestable priority by the state, massive research institutes established under Stalin to benefit industries showed an interesting tendency to fundamentalization. While their raison d'tre was to introduce technological innovation into the industry that it was attached to, neither the industry nor the research institutes were motivated to do so. The risk of disrupting industrial production as a prelude to innovation was real indeed while the benefits that innovation might bring were illusory at best. The managers of industrial enterprises would shy away from innovation. Directors of research institutes were among the first to acknowledge this attitude and had to readjust their priorities to ensure their institutes' survival and growth. It would be risky to predicate their success on that of actual industrial innovation over which they, the researchers, had little control. Gradually, applied research institutes realigned their activities in such a manner as to maximize the output of doctoral theses, scientific publications and other usual products of fundamental research.

Adroitly using Marxian words about science as a direct productive force, they argued that such scientific achievements would necessarily percolate into industry and improve its performance.

This cultural change began in the mid-1930s but really took root under Khrushchev and Brezhnev. It strengthened the ethos of science-driven innovation that left post-Soviet republics, and above all Russia, with an impressive R&D potential but few if any skills to market new ideas to those who could turn them into new processes and products.

The Soviet culture of innovation left scientists and engineers unprepared to face new conditions after the collapse of the USSR. While many a scientist used their well trained minds to attain spectacular business achievements, few of these had much to do with science or technology.

To plan science in research institutes was commonplace. However, the efficiency of such planning was dubious as industrial plants, in theory the ultimate consumer of innovative ideas, were rarely interested in implementing them. Industrial consulting done by research institutes was common throughout most of the Soviet years. This not only brought extra income to the institute and its scientists, but enabled a freer interaction between science and industry than that established through central planning.

Applied scientists devoted larger share of their time to organizational and paper work than their colleagues engaged in fundamental research.

The establishment of a special academic city near Novosibirsk was a major event of Soviet science policy. Genuine enthusiasm and relative freedom in Akademgorodok enabled Siberian scientists to develop a particular ethos which accounted for their remarkable scientific success in the 1960s and 70s. However, Akademgorodok failed to satisfy the government expectations in terms of advancing industrial technologies.

Attempts to innovate in the diffusion of innovations (the experience of Fakel, a local innovation firm) were met with opposition in both Moscow and Siberia party circles,

and produced little in terms of transfer of technologies to industry. Cultural and political barriers prevented the emergence of a viable innovation system. At the same time, the ethos of fundamental research got only stronger.
It was commonly beleived that rather than putting emphasis on the uses of science, it should rather be viewed as an intrinsic value, as "part and parcel of modern culture". This view was reiterated by the Rector of an elite educational establishment, a major supplier of S&T personnel for the country's nuclear, missile and other advanced technology programmes: "The fact that one cannot manipulate science constitutes the main socio-psychological and administrative-financial difficulty. By its very nature science, particularly basic science, develops according to its own logic. One can do nothing about it. Either there is science, or there is none. And if it exists, it must be financed decently. Society must support science as such, without requesting immediate useful yield, but rather expecting as an indefinite prospect..." These views continued to be expressed even in the early post-Soviet period when the country's R&D system underwent rapid attrition.

4. SOVIET INNOVATION PERFORMANCE

The Soviet concept for technological innovation whether in civilian or military spheres was vnedreniye. This word means insertion, imposition of an alien entity, it also implies resistence to this intervention. Peter Kapitsa, one of the most prominent Soviet physicists with an extensive British background, remarked that the term vnedreniye stands for "abnormal conditions of assimilation of new technology".
"When we begin to use the term assimilation (osvoyeniye), he continues, one could then say that we have reached normal conditions for its [new technology's] development." However, the term was consacrated in the Soviet Constitution (article 26): "The State, in accordance with the demand of society, organizes vnedreniye of results of scientific and technical research into the people's economy and other spheres of life." Vnedreniye had many adepts, particularly among experts in charge of transmission of inventive ideas into practical use. They saw vnedreniye as a process spanning the entire spectrum of activities, from basic research to industrial production.
Vnedreniye of new technologies was of acute concern to political leaders throughout Soviet history. One finds unintentionally prophetic words in the proceedings of the 26th congress of the Communist Party which called "to sweep away all that makes the process of vnedreniye difficult, slow, painful". Indeed, when the Soviet economic system was swept away this reanimated optimism in greater use of scientific and technological achievements. Significantly, by the end of the 1980s, the term vnedreniye had virtually disaapeared from the vocabulary of Soviet specialists in innovation studies.
The Soviet Union is known to have exported advanced technologies, usually in the form of patents to industrialized countries, and in the form of equipment and even turn-key factories to developing nations. A book published in 1989 in New York praised "unknown innovators in the global economy". The heroes of the book were citizens of Eastern Europe, including Russia, who, "as the inventors have to become reformers in

order to realize their ambitions." The author John Kiser shows how Soviet, Hungarian and East German technologies were successfully transfered to major industrial companies in Western Europe, North America and East Asia. He argues for more appreciation of Soviet bloc technologies and shows how this appreciation pays off. The author claims to head one of the very few profitable companies in the world based on import and diffusion of Eastern European technologies.

Emigration from the USSR, another indicator of innovative activity, included specialists who became leaders of several technological fields in the new country. For example, I I Sikorsky or V K Zworykin, attained fame as talented engineers and inventors, associated with helicopters and television, respectively. George Gamow became known for his work on atomic nuclei and the properties of elementary particles. Another Russian scientist, Vladimir Ipatieff, an eminent chemist who arrived in the United States at the age of 62, brought with him unique expertise in catalytic organic chemistry which he successfully introduced into American petroleum industry. During his two decades in the United States Ipatieff authored scores of scientific papers and industrial patents, and trained dozens of specialists in catalytic petroleum chemistry. Several decades later, Soviet chemists concerned about the decline of the science of chemical catalysis in their country would seek Ipatieff's American disciples in the framework of Soviet-American scientific exchanges.

The fact that the Soviet Union could export patents and experts to most advanced industrial companies in Japan, Europe and North America cannot obliterate its profound dependence on technology transfer from the West. Arguments to the contrary as so overtly ideological that they hardly deserve refutation.

Culturally conditioned perceptions in the course of the Cold War did exaggerate the technological backwardness of the Soviet Union. This accounts for the surprise that the launching of the Sputnik in 1957 constituted for Western, particularly American observers. They had then a conceptual difficulty to understand Soviet successes in technological innovation in narrowly defined priority fields. It was hard to fathom a society that could produce the world's best rocketry but had to import plumbing equipment for its tourist hotels. This conceptual difficulty was caused by the habitual assumption that technologies tend to get diffused within a given economic space so that the technological levels of all the industries are more or less in harmony. However, such diffusion of technologicsal innovation should not be taken for granted, and is in itself culturally conditioned.

By the end of the 20th century several countries exhibit pronounced disparities between the technological development of certain, usually defense-related sectors and the rest of the industrial plant. The priority sectors absorb the bulk of available capital and specialized human resources leaving the rest of the economy in its primitive state.

The most drastic concentration of resources of this nature can only be accomplished in totalitarian societies such as the Soviet Union, North Korea or Communist China. However, the examples of India and Pakistan suggest that governments in relatively democractic countries can also allocate massive resources to science-based industrial development. This happens in spite of the fact that such investment may starve the rest of the economy of capital. Hoped-for spillover effects on industries left decades,

sometimes centuries, behind in technological sophistication rarely maerialize in such countries.
Innovation is essentially a disruptive and risky process. One has to have very good reasons in order to embark on innovative strategies. In market economies, the fear of losing out to the competition and the prospect of dramatically augmenting profits combine to produce an environment for diffusion of new technologies. In the Soviet Union, neither of these economic factors were ever operative. Rather, political factors and the basic physical survival instincts were activated instead.
Most successful cases of innovation in the Soviet Union are tributary to direct involvement of the political leadership at the highest level. Stalin, Beria, Khrushchev assumed direct responsibility for projects such as the development of new aircraft, the creation of the Soviet nuclear weapons or the production of ballistic missiles. Not only their personal involvement removed bottlenecks characteristic of the rest of Soviet economy, it also instilled fear in scientists and engineers who found themselves exposed to certain retribution in the case of failure.
Best cases of technological innovation involved the military that enjoyed practically unlimited resources. Andrei Sakharov makes an explicit reference to this culture of innovation in high-priority projects when discussing the events leading to the first thermonuclear explosion. These projects brought together high-calibre scientists, engineers and industrial managers who engaged in genuine cooperation. This cooperation was unusual in a country known for the insurmountability of barriers among competing institutions, albeit nominally subordinated to the state. This close cooperation and the lavish spending that it was supported with, created a unique culture that eventually went beyond the confines of the nuclear programme.
After the deaths of Stalin and Beria in 1953, and the concomittant removal of the terror, physicists became aware of their privileged position as perhaps the only professional group that the state found indispensable. Massive efforts at technological innovation, such as the nuclear arms programme, produced major dislocations not only in the economic but also in the political sphere. While the A and H bombs consecrated the Soviet Union as one of the two superpowers, seemingly impermeable to political change, the bombs eroded the Communist monopoly of political control and, ultimately, the solidity of the state institutions. This led to the emergence of dissidence that in turn contributed to the end of the Soviet regime.
A privileged tool of technological progress under Stalin was an engineering office confined to a concentration camp (sharaga). Prominent engineers and scientists were arrested and made to work in sharagas specializing in a wide variety of advanced technologies. David Holloway, an expert in Soviet military innovation, characterizes "the use of prisoners, condemned as enemies of the state, to design and develop weapons for the defense of that state" as "one of the most bizarre elements of the NKVD empire". Tupolev, one of the founders of the Soviet jet industry, and Korolev who spearheaded the missile development in the Soviet Union, spent years as prisoners in, and at the same time, intellectual leaders of, their respective sharagas. Solzhenitsyn's The First Circle describes how one such sharaga was developing a voice identification device for the secret police. Beyond military projects, sharagas worked on a variety of

civilian technologies such as design, in 1932, of a new powerful locomotive appropriately named "Iosif Stalin".
The recruitment of German nuclear scientists for work in the Soviet Union came to light in the wake of glasnost. Sakharov mentions German expatriates briefly, and inconclusively, in his memoirs. Later it became known that German scientists were organized in two research institutes located on the Black Sea coast of Georgia. Their results were channeled to Moscow for revaluation and development. Some German scientists were allowed to take part in industrial application of their ideas, namely in the transpolar city of Norilsk. They were invariably treated with respect and accorded privilege customary among the higher echelons of the Soviet science system. In a telling anecdote of the last days of the Second World War, one of the scientists who would end up working in the Soviet Union, was visited by a Soviet chemist in military uniform who placed a Russian-language sign on the German's door: "Here lives a senior scientist. Do not disturb." In the meantime, the massive attack of the Soviet army on Berlin went on. One German sceientist took with him a train load of laboratory equipment, including a cyclotrone.
The actual contribution of German scientists to the Soviet nuclear programme is hard to guage but it appears nowhere close to the role they played in the development of weapons systems in the United States. One factor that may have limited their contribution is the severe control and compartimentalization of information which was enforced in the Soviet nuclear project, and a fortiori with respect to German expatriates. Such information controls were endemic in Soviet science which operated under the presumption that all scientific data were secret unless certified otherwise. While admitting that defense research was conducted likewise in many countries, a senior science manager wrote in 1993 that "the peculiarity of our science which consists in an a priori consideration of any new scientific field as a (military) secret; one must subsequently prove that it is not secret, even though there is no factual or nominal belonging of this field to the military."
This behaviour did substantial harm to Soviet technological progress. Western sceptics had long attributed the rapid production of the Soviet A-bomb to Soviet espionage in the United States. They found a new ally in General Sudoplatov, former senior intelligence officer, who recounted his side of the spy story in 1994 and, not unexpectedly, reinforced the belief in the omnipotence of Soviet intelligence. The question of the origins of Soviet nuclear weapons remains wide open, and may remain so indefinitely.

5. STAR WARS

President Reagan made the first announcement about the Strategic Defense Initiative (SDI) in 1983. It was soon dubbed Star Wars and provoked substantial dissent among American scientists. The feasibility of the scheme that would hermetically protect the United States from incoming nuclear weapons was often put in doubt, and it has remained a divisive, albeit nowadays an historical, subject since its inception.
However, the most important effect of the SDI was produced far from the American

shores, namely in the Soviet Union.
The SDI was rightly perceived as a serious threat to the parity of terror between the two superpowers. Unless counteracted, it would have made ineffective and therefore useless the impressive nuclear stockpiles of the Soviet armed forces. Soviet scientists and engineers were mobilized to study the SDI project and to come up with a counterinvention. However, President Reagan's announcement hardly took Soviet scientists and military strategists by surprise.
Soviet scientists began working on laser weapons as early as 1960. While future Nobel laureates Alexander Prokhorov and Nikolai Basov presented their laser at the Academy's Lebedev Institute in Moscow, secret work was underway at the State Optical Institute (GOI) in Leningrad. Dmitrii Ustinov, then in charge of military procurements at the CPSU Central Committee and later the Soviet Defense Minister, took a strong personal interest in the military applications of the laser. He would later appoint his son Nikolai to lead the development of the new weapon.
Academician Gersh Budker made a first proposal to build an antimissile system in space in the 1960s but his idea, in spite of the spectacular Soviet successes in space exploration, failed to find support in the Kremlin. However, in 1975, eight years prior to Reagan's announcement, Soviets began exploring an antimissile space shield. They concluded, by the end of the decade, that a system capable of destroying 10 000 warheads in the lapse of 5 to 25 minutes, was not feasable.
According to post-Soviet evidence, several countermeasures to the SDI were developed in largely intellectual terms. One of the most imaginative was a "dissymetric response" proposed by Anatoli Savin who had himself directed Soviet research on the subject in the 1970s. His idea consisted in improving the number and the penetrating capacity of Soviet missiles, and its realization would cost about one hundred times less than the American SDI. Once accepted, his programme would have increased the Soviet military expenses by about 7%. However, even this relatively modest response to the challenge posed by the SDI would have strained Soviet resources beyond limits acceptable under Gorbachev.
A broad front of research would mobilize thousands of scientists across the Soviet Union. The Academy of Sciences and other "civilian" institutions became intimately involved in the military programme
alongside with the traditional secret research centres such as Arzamas-16 or Tomsk-7. The programme developed a few successful devices. In 1984 Nikolai Ustinov attended a real-life laser targeting of the U.S. space shuttle Challenger that the Soviets believed to be primarily a military vehicle. It was a sign of the changing times, that another successful device, the Topaz 2 nuclear reactor for use in space, would be sold in 1993 to the U.S. SDI project. Like the Americans, the Soviets did succeed in destroying a tactical missile in the air but the innovative equipment was far from being operational.
In 1985 the new and relatively youthful leader Mikhail Gorbachev paid personal visits to several institutions working on the Soviet response to the SDI. He renewed the State's commitment to the development of laser weapons, and his scientific advisor Academician Evgenii Velikhov would disburse ca. $100 million a year to support R&D in this field.

Within a few years Gorbachev realized the financial magnitude of the SDI programme, and his country's impossibility to meet the American challenge. After a prolonged period of denouncing the SDI, he began to look for accomodation with the United States. This opened new prospects in the relations between the two superowers which acquired a progressively cooperative nature in the late 1980s.
Some argued that the SDI triggered the collapse of the Soviet system. "Gorbachev's response to Reagan's challenge ultimately destroyed Communism." Indeed, Gorbachev had to commit massive resources of the 1986-90 five-year plan to match the US programme. His original focus on technological innovation was real. However, he soon realized that military competition with the West was neither feasible nor affordable.
Soviet generals' "determination to have a full-blown strategic defense of their own dramatized the internal price of the arms race and made clear the need to shake the system up."
This is how military imperatives, which had constituted the main reason for the development of Russia's and Soviet science, destabilized the Soviet state before finally losing their importance. The disparity between the intellectual and material resources of the Soviet Union, made clear by the SDI episode, underlined the vulnerability of Soviet society and, at the same time, the precarity of the immense science and technology system in the USSR. The Soviets' familiarity with the scientific background of the SDI was an important part of the American project one of whose goals, and certainly not the least significant, was to destabilize the economy and ultimately the political system of the Soviet Union. The SDI is often credited with moving Gorbachev to seek Western approaches to repairing the economic state of the country. It is assumed that this ultimately led to the demise of the Soviet Union and the virtual collapse of its economy.
The SDI also accelerated the crisis in Soviet science by further eroding the nomenklatura's traditional belief in the omnipotence of science and technology, in their ability to fix the country's economic and military problems. Science was growingly perceived as an expensive luxury and, with the weakening of the old Communist ideology, the state's commitment to science was drastically revised downwards. The SDI turned out to be a crucial catalyst in this process. The impressive R&D system built to develop Soviet laser weapons began to suffer from shrinking funds in the late 1980s. The abortive coup d'tat in August 1991 turned out to be a coup-de-grce not only to the Soviet Union, but to the Soviet laser weapon project. It began to disintegrate, with a few offshoots attempting conversion to civilian use. The Government of Russia tried to stop the disintegration but the exodus of the most qualified personnel eroded the intellectual potential of the programme beyond repair.

6. POST-SOVIET DEVELOPMENTS

Soviet science had no independent sources of financing, and generally lacked formal horizontal networks of communication. This sociometric pattern was enforced by the Soviet regime which discouraged independent associations, tended to atomize society and render its members politically impotent. In a way science, because of its intrinsic collegiality, has been shown to be less affected by the dominant atomization than other sectors of activity. However, in terms of financing, the role of the administrative command system was undisputed, and the few attempts to introduce khozraschet and other forms of contractual R&D changed little in the overall monopolistic picture.

Nauchno-proizvodstvennye ob'edineniya (research and production associations) which had been introduced in the last two decades of the Soviet Union apparently did little to make R&D more relevant to the needs of industry.

In the wake of the collapse of the USSR, government is no longer considered a reliable agent of technological change. Thus a legislator intimately acquainted with S&T policy making and implementation wrote in 1993 that exclusive reliance on a government policy was "hopeless".

"Conversely, private capital, interested in profits and efficiency of production, automatically strives to modernize production, to attract new ideas, new technologies, and new people capable of managing production under new conditions."

The belief in the virtues of capitalism may appear somewhat ideological: it reflected a high degree of consensus, possibly borne out of frustration, that emerged after 70 years of anti-capitalist rhetoric.

Current problems of Russia's S&T are due to its Soviet heritage rather than solely to post-Soviet developments. It has been remarked that the links between R&D and manufacturing enterprises were never really important . Otherwise, the industry would not have found itself in as acute a crisis as it did. The weakness of these links was simply exposed and became obvious in the post-Soviet economic crisis. Neither internal nor international markets were interested in absorbing significant amounts of technology-intensive products made in Russia. Consequently, there is little need in the big science created by the USSR in response to its politico-military interests and ideological ambitions. Applied research and development work conducted in Russia has departed from the Soviet pattern of producing "paper" of little relevance to the customer.

The Soviet industrial system put ritual emphasis on technological innovation but, with a few exceptions, abhorred disrupting and risky ventures which innovation by definition implies. Nowadays, some remnants of the old system have come to respond to the customers' micro-problems (find substitution for missing raw materials or devise modes to produce parts no longer available in Russia). Usually this happens in industries that have entered into joint production agreements with Western companies. Significantly, little work is reported done on more than incremental problems, on ways of eliminating immediate bottlenecks.

The Soviet culture of innovation left scientists and engineers unprepared to face new conditions. While many a scientist used their well trained minds to attain spectacular business achivements, few of these had much to do with science or technology. Some experts consider the crisis of conversion to be of systemic rather than structural nature: it is not enough to switch production from submarines to leisure boats, one must also learn to find markets for its products. However, managers of state sector enterprises believe there exists no demand for R&D in the state sector. Their strong support for R&D may be due to the expectation that industry will finally turn around and will have to need R&D. It is significant that in 1993 the managers expressed no specific

expectation as to actual contributions of R&D to their enterprises.
In accordance with the traditional belief in supply (i.e. science), rather than demand generated innovations, many scientists and engineers continue to lament that much of the country's R&D remains "unsolicited", even after the colapse of the USSR. They are upset that scientific ideas, some claimed to be"of planetary scale" such as formulated by Vernadsky or Dokuchaev, have found no practical application. "There are hundred of thousands of technical documents, particularly patents, that lie like auriferous ore with low concentration of gold".
This somewhat simplistic view of post-conversion innovation and production opportunities is rather common. For example, the director of a formerly defense-oriented R&D institute laments that it cannot find a manufacturer for a complex piece of diagnostic equipment; it remains unclear, however, if there would be a market for such equipment in Russia or outside. The head of another institution that had designed bacteriological weapons complains that the break-up of the USSR deprived his institution of opportunities to materialize their ideas in production.
There is currently little expectation that, in the short term, the emerging private sector would need S&T, even though many of the new entrepreneurs hold advanced scientific degrees. "Under present circumstances, every company tries to invest its profits in the most advantageous manner. Long-term venture investments which may, or may not, bring substantial dividends in an indefinite number of years are not attractive. From this point of view, even technological innovation is too risky; investment in purely scientific developments is utterly senseless."
Moreover, it is important in cultural terms that there emerged the following consensus: "society no longer sees in science the main source of material and moral values, it no longer deems science to be among the most honourable or useful activities". Consequently, some observers perceive that the government considers it "unprofitable" to support science. Scientists came to earn 60% of the average wages compared to 115% in the last years of the USSR. According to a prominent member of Russia's legislature, "No party, no political movement understands science. Science is usually seen as a parasite." Moreover, many peasants resent agricultural science. Politically motivated decisions used to be imposed on them in the guise of "scientific views" under "the Agrogulag", the system of centralized administration of agriculture. In the eyes of the peasant, as reported by the head of the Peasants' Party, the distant arrogant science was the habitual scapegoat for poor harvests and all their misfortunes.
The falling wages are by far not the only, or even the main source of difficulties for S&T in the new Russia. While public confidence in science, education and scientific planning has been undermined, the interest has been revived in "mysticism, occultism, and archaic world views."A part of the bureaucracy and intellectuals have rushed away from rationalism and scientific cognition of the world", laments a well-known organizer of science from the South of Russia. The abuse of conscience manipulation in the Soviet period "has brought the country to a medieval stage and occult mystical beliefs. Scientific approach is alien for the simple Philistine...Very few actually revere science."
The precipitous decline in the status of science in the last decade has extinguished expectations of science to become a crucial factor in the new liberal and democratic

world. However, it did not become a scapegoat, the source of all evil. Its relevance has simply shrunk. There have been no pronounced anti-science views or organizations reported in Russia's mainstream media.
A survey on public attitudes to science revealed that only 8% of the respondents believed that "science did more harm than good", while 52% saw more good than harm in science. In the ascending order, the harm
done by different sciences appears as follows: medical sciences, biology, physics, humanities, social sciences and, finally, chemistry - the most maligned of the sciences. The country's difficulties are blamed "on politicians rather than on scientists" by two thirds of the respondents.
Most respondents expect science to yield immediately useful results, while only 14% support basic research. Not surprisingly, the least educated strata exhibit less enthusiastic attitudes to science. While many industrial managers are willing to donate a part of their revenues to support science, very few peasants and unqualified labourers share this inclination. The legislative branch of the government appears less supportive of science that the executive branch.
It is believed among some post-Soviet elites, that significant weakening of the intellectual and scientific stratum in Russia would lead to a rapid political shift to the right in post-Soviet society in general. The breakdown of the intellectual class would reinforce the society's "aggressive tendencies", and thus increase the risk of an international conflict. This argument, occasionally advanced by politically active scientists before, is shared by an historian, head of a major university and a prominent pro-reform politician . A political activist from the Academy of Sciences, remarks that "the democratic government of Russia committed a political error having neglected to develop a safety net for scientists. It thus has cut off the branch on which it was sitting. Khrushchev had made a similar mistake; and this played a non-negligible role in the failure of his reforms and was taken into account by Gorbachev, which enabled him to operate a bloodless anti-Communist coup d'tat." Furthermore, "an outflow of scientific elites will lead to a lowering of the country's intellectual level and will facilitate a return to the Communist past." One also reads that "the disappearance of scientists in Russia must cause no lesser concern that that of any species listed in the Red Book [which lists species threatened with extinction]." "A generation or two of young people removed from the process of transmission of knowledge can effectively arrest this process on one sixth of the planet."
The place of science and technology in the rapidly evolving cultures of post-Soviet Russia is clearly becoming more modest. This is a reaction to Soviet scientism and the apparently intimate links between science and Communist ideology. These links were often translated into significant privilege and exceedingly generous, by Western standards, allocation of resources to science and technology. In spite of the resentment this has caused after the fall of the USSR, no anti-science movements have appeared in Russia, and that few question the importance of science, be it intrinsic or extrinsic.
The morality of science acquired a great importance under the communists. In conjunction with pleas for a more humanized S&T in the wake of the break-up of the USSR, one also encounters ardent calls for repentance, coming, significantly, not from

a cleric but from an applied scientist. "In order to put into motion the mechanism of internal self-cleaning and rehabilitation, our science needs a merciless and uncompromising self-analysis and repentance for past sins. In many respects, these sins, quite common for both science and higher education in view of their genetic unity, consist in the amorality of many of our professional occupations, paid for by the toil of the people, but practically useless and even harmful to the people." The scientist, whose opinion was solicited among others in a survey conducted in 1993, brought examples of militarization of technology while "the everyday needs" were routinely neglected, and consumer technologies were usually transferred from the West with considerable delays. In this light, one may be surprised that a strong commitment to science was articulated by a prominent cleric of the Russian Orthodox Church (whose lectures used to draw hundreds of scientists in the initial stages of glasnost): "If science can accept as its basis the primacy of spiritual and ethical values, and be able to combine the material and the spiritual, it will have a good prospect for the future." In other words, scientists should part with their autonomous science-based morality which they used to extend so willingly to many other domains.

7. RESCUING RUSSIA'S INNOVATION POTENTIAL

"Nowadays, we are forced to play the role of a gigantic gas station in the global European exchange of goods; we should fight to be eventually recognized as a supplier of innovations", was the common leitmotif in the early 1990s. Novel and seemingly practical ways of providing foreign assistance to the S&T system were debated in Russia's press. For example, a businessman with a background in R&D suggested that beyond providing grants to individual scientists and research groups, the international scientific community, rather than governmental agencies, should set up a non-profit corporation which would channel R&D work for execution in Russia and would transfer Russian technologies abroad. The obvious economic advantage is the low cost of Russia's R&D.

The corporation would serve as a monitor, a clearing house and, above all, an intermediary between Russia's laboratories and foreign firms, foundations, and governments. Operating through a Western bank with significant experience in Russia, the corporation would release funds in stages, according to the progress of a given project. However, the allocation of these funds should be under the control of the foreign-based corporation rather than Russia's administrators of science and technology. A variant of this idea is the expectation that Western businesses would integrate Russia's R&D into its worldwide network since investment in science is deemed to carry "virtually no risk compared to investment into industrial production." This assessment becomes, however, less optimistic with respect to investment on the part of Russia's nascent private sector. Moreover, rapid advances in instrumentation make it unlikely that Russia's scientists and engineers would actually be assigned labour-intensive research or mere routine work.

In technology and applied research, international cooperation is obviously conditioned by the advances in economic integration of Russia into the world economy. The removal of secrecy restrictions by Russia, and of COCOM regulations by the Western world has opened new ways for cooperation. In the field of computers, where Russia is clearly behind, there can be observed a move from old-style piracy of Western prototypes to joint research and development. Russia's traditional reliance on technology pirated from certain Western corporations, such as Digital Equipment, has already led to cooperative research with the private sector in Russia. Soviet scientists' skills and imagination have been sharpened by years of the uphill battle to compensate the West's rapid advances in hardware with original software solutions. Other fields of science have also displayed this comparative advantage of post-Soviet researchers to resort to inexpensive, unorthodox methods in the conditions of scarcity of scientific instrumentation and to arrive at high-quality scientific results. "This skill of Soviet scientists may be no doubt useful for Western science."

"Foreign investment should not be used to support clumsy and superfluous state institutions, but rather to stimulate science in the private sector" was a frequent opinion in the early 1990s. It was also suggested that foreign capital and marketing skills be used to organize a transfer of Russia's technologies, reputed to be inexpensive, ingenious and less hardware dependent, to countries in the Third world.

8. CONCLUDING REMARKS

Russia's conservative circles often express a concern that there has been a substantial drain on the country's reservoir of innovative ideas, accumulated in Soviet state institutions but usually sold abroad for narrow personal gain. Outbound transfers of important technologies are sometimes deemed to produce no equivalent benefits for the nation ("diamonds are exchanged for glass beads"). This claim is predicated on the science-driven model of industrial innovation. It is occasionally repeated in the country's nationalist media, so far with little empirical evidence as to the actual economic benefits derived by a Western company from Soviet know-how. At the same time, it is commonly accepted that the share of intellectual product in Russia's exports is likely to remain low for several years.

The tension between pro-Western and anti-Western ideologies of Russia's development is relatively less manifest in the sphere of S&T.

Some scientists and even science policy experts may take extreme ideological positions on issues of foreign policy or private ownership of land but, significantly, become more consensual when approaching the future of S&T, presumably the domain they should be more familiar with. It follows that mutually profitable international cooperation in S&T may be easier to organize than cooperation with the West in other spheres.

Scientists are more acquainted with the culture of world science than industrial managers or military officers with respective cultures obtaining in the West. S&T may serve as a good springboard and a convincing example for equitable cooperation in other more sensitive matters. A note of warning is sounded with respect to import of science- and labour-intensive technologies into Russia. Such import is deemed not only useless but potentially harmful. It would strengthen the nation's technological dependence, and possibly lead to a proliferation of environmentally hazardous industries which are being dismantled in most industrial democracies. Moreover, it would inhibit adaptation of the "gold reserve of unsolicited technologies" to civilian needs.

The concept of "unsolicited technologies" remains an important vestige of the Soviet innovation culture. It is intimately related to another key concept, that of vnedreniye which may remain popular in the current generation of technology managers. Together they bring back images of forced modernization, culturally rootless ("hydroponic") science and technology, and uneasy public attitudes to the religion of engineering progress that used to dominate the Soviet scene. These cultural obstacles to innovation must be kept in mind alongside the more obvious political and economic factors.

TENDENCIES OF DEVELOPMENT OF SMALL INNOVATIVE ENTREPRENEURSHIP IN RUSSIA

BORTNIK I.M.
General Director of Fund for Promotion of SMEs in Scientific and Technical Area.
Moscow, Russia

KOZLOV G.V.
Deputy Minister of Science and Technical Policy of Russian Federation.
Moscow, Russia

Two objective reasons, both caused by the change of political and economic situation in Russia, were powerful factors of development of SMEs in Russia (Fig.1).

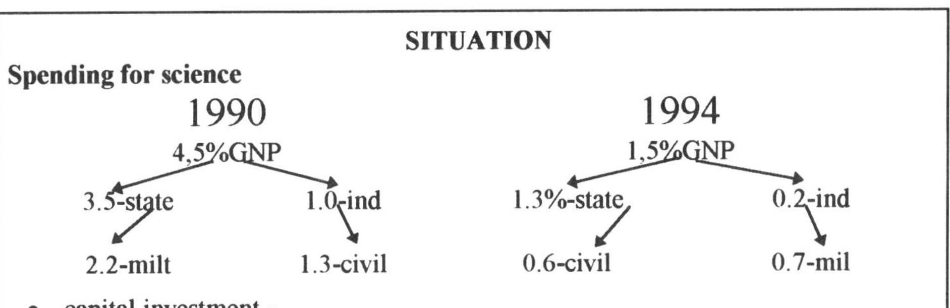

- capital investment -
- production -
- imposted goods -
- science is not only area for ambitious people, other areas are open
- money are now much more important for life
- money eases to make in other atlas.

Fig. 1

The first one was concession of freedom to create independent firms. What's more, this process was further stimulated by concessional taxation (tax vacation) for SMEs in the number of priority areas, innovative in particular.

The second one was sharp reduction, practically by 4 times in real figures, of financing of Russian science even from federal budget, which always financed fundamental science, state scientific and technical priorities, military science. Because of sharp and simultaneous fall in production and capital investments, the industry practically stopped financing of applied science. This figure was always approximately one half of the total financing of the civilian science.

In total reduction of scientists in Russian, the subjective factor plays an important role. A number of "unworthy" professions for young men in earlier years, such as

C. Corsi (ed.), Science and Innovation as Strategic Tools for Industrial and Economic Growth, 51-56.

businessman, accountant, financier, clerk in banks, became prestigious and considerably highly paid in addition to it. The money itself became much more an important life factor.

The whole complex of above-mentioned reasons and factors has led to considerable reduction of 4000 research institutes and scientific organizations, existing in Russia in 1990, from 3 millions of employers to 2,2 millions. At the same time more than 60 thousands of small innovative enterprises with a global personnel of 400 thousands of people were created (Fig.2).

SE - INNOVATIVE - 1993

ENTERPR.	**EMPLOCES**	**INVOLVED E MPL**
60.000	400.000	2.000.000
PRIVATE 90%	80%	

REGIONAL DISTRIBUTION

MOSCOW & REGION	**ST-PETECSB & REGION**	**VOLGA REGION**	**URAL**	**SIBERIA**
50%	20%	7.3%	7.3%	10%

PROBLEM: REAL STATISTICS

Fig. 2

It should be mentioned that distribution of the quantity of SMEs in the region fully reflects the distribution of scientific and technical potential in Russia; in (Fig.3) the result of the first official statistic analysis of SMEs in Russia in 1993 is presented.

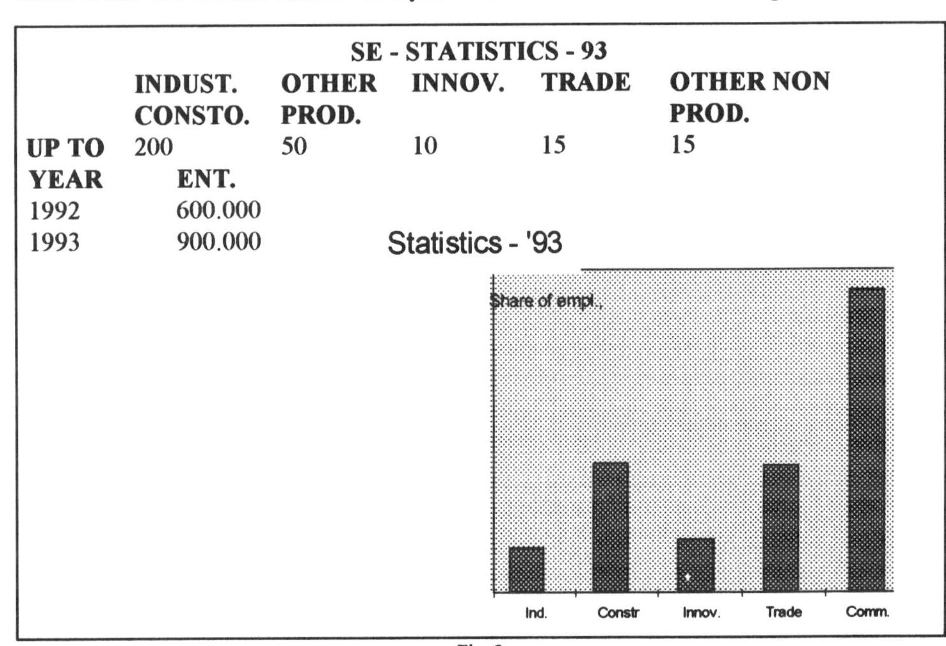

SE - STATISTICS - 93

	INDUST. CONSTO.	**OTHER PROD.**	**INNOV.**	**TRADE**	**OTHER NON PROD.**
UP TO	200	50	10	15	15

YEAR	**ENT.**
1992	600.000
1993	900.000

Fig. 3

This statistics is imperfect yet and criteria of SMEs changed after the federal Law on "Governmental Support of Small Entrepreneurship in Russian Federation" was adopted in June 1995.
Actually on enterprise is considered "small" if its personnel are not more than 100, in industry and 60, in scientific and technical area, moreover it is to be underlined that 16% of all employees in the area of science and scientific services work in SMEs. The actual tendencies in economic situation in Russia are shown on (Fig.4.)

TENDENCIES

- capital is supplied by some scientists & engineers (through commerce)
- "angels" appears
- not so easy money are now in commerce and finance areas (no safety)
- price to product is high enough to use imported element
- still low labour cost.
- new factories'owners (financial capital) are looking for profit, i.e. new products, technologies, managers
- financial structures start to search for high technologies areas to invest (informatics, telecommunication)
- there is some tax relief for SMEs
- new companies (also foreign) are looking for local product and subcontractors
- there is now Federal program for SMEs support
- new legislation for SMEs is prepared

Fig. 4

At the same time in the course of its development and activity SME experiences varions difficulties, shown on (Fig.5).

PROBLEMS

- still bad credit situation for SMEs
- foreign product (open market)
- low level technology
- low quality infrastructure
- bad information
- small buying capacity
- weak market for subcontract
- low level of personal income in general and corresponding taxation scale

Fig. 5

Some of them are characteristic for all SMEs, the others are specific for Russian SMEs. A main difficulting concerns, in particular, the problem of intellectual property. Infact, in many cases this problem is not worket out between SME (though inventors work there, as a rule) and former copyright owner, normally a big company.
The process of governmental support to small entrepreneurship in Russia is on the

initial stage (Fig.6).

FUND OF ASSISTANCE TO INNOVATIVE SMES
(FEDERAL, 0.5% OF FEDERAL SCIENCE BUDGET, 1994 - 25.0 BLN

Created: 03.02.94

Situation 25.10.94:
- 6.0 Bln in Rbl
- 6.0 projects
- 13 Regions of Russia
- Usual expertise system
- 10% financing 50/50 bases (return in 1 year)

Projects:
- R&D
- initial production
- training
- patenting
- certification
- conferences' support, publishing

Fig. 6

As far as legislative assurance is concerned besides above mentioned "tax vacation" for SMEs, acting in a number of priority areas, the Law of 1995 stipulates simplified order of registration and strategical accountance of SMEs. The system of funds for support of small entrepreneurship with participation of federal and regional budgets, having the rights to preferential crediting of SMEs is being formed. Besides budgetary resources these funds receive part of income, getting from privatisation of State property.
The State also participates in formation of substantial part of the market for SMEs production. The Law determines that not less than 15% of State order should be placed at SMEs.
The Law stipulates considerable increase of admissible norms of writing down depreciation charges on production costs.
The draft of the Law on simplified system of taxation of SMEs was prepared and adopted by the State Duma. It also determines the limit of total value of taxation for SMEs.
The Decree of the Government on "Development of Leasing in investment activity", which must considerably facilitate the financial burden of SMEs was adopted in June 1995.
For the purpose of creation of organisational structure for promotion of small entrepreneurship, by the Decree of the President of Russian Federation, the State Committee for Small Entrepreneurship Support was formed; the activity of different state and regional organs is coordinated in the framework of the Federal Programme for the Development of the Entrepreneurship, which stipulates elaboration of the legislative, normative and juridical basis, development of infrastructure for entrepreneurship support, as well as specialized information network and resource base,

support of concrete projects of small enterprises.

As regards the support of small innovative entrepreneurship, in the beginning of 1994 the Government of Russian Federation founded the Fund for Promotion of SMEs in Scientific and Technical Area. The financial basis of the Fund were annual deduction, equal to 0,5% of the Federal budget, stipulated for science (Fig.6). The main objectives of the Fund are as follows: to develop more effective and revisable mechanism of state support of the process of formation of stable small enterprises, developing and realising products and services on the basis of intellectual property, belonging to SMEs in the conditions of market economy. The Fund as occupied it's place within the system of Funds of Russia, supporting development of science and technology (Fig.7).

FUNDS

Federal

- basic research
- humanitarian research
- to support innovative SMEs
- to support enterprenership and competition
- technology development

Private (started with states)

- Petrov
- Scerbakov
- Nechaev

Ministry's level to support industries
Public Organizations (funds, unions, associations, etc)
5% of Fed. budget, 5% of privatization, 1,5% of realization.

Fig. 7

For one year and one half the Fund have been mastering different mechanisms of support for the development of SMEs. The indices of the initial activity of the Fund are shown on Fig.6. Nowadays the Fund supports more than 150 projects in 24 regions of Russian Federation with total value of 6,0 millions dollars.

The analysis of activities of the SMEs supported by the Fund shows the following: SMEs can't exist, manufacturing products which are not in demand. Therefore they orient on development hand manufacturing of the products, which have its payable buyer. The principle buyer is population nowadays. That's why the production of SMEs is socially-oriented, in general. (Fig.8).

ATTRACTIVE FIELDS FOR INNOVATIVE SMES
(where people buy and pay)

- medicine
- home ecology (clean water)
- personal safety
- software and hardware
- packing
- industrial ecology and monitoring
- office product
- energy saving

Fig. 8

The process of formation SMEs is under way:

- the average turnover of SMEs nowadays is approximately 10 thousands dollars per year; its stable functioning is observed at turnover on the level of 500 thousands dollars per year:
- overwhelming majority of SMEs functions thanks to the fact that they rent spaces, research, test and production equipment of state enterprises (or former state enterprises, which don't start active functioning yet) free of charge, or for very small payment;
- the major problem of SMEs is attraction of financial resources for its development and current assets inflow under conditions of lack of considerable capital assets, active reserves and small turnover. In this situation they can make loans only if interest rate is equal to 1/5 - 1/4 of the Central Bank discount rate, under liberal conditions of return and mortgage requirements of the creditor;
- matters of accurately drawn up rights of intellectual property and its protection are not yet of vital importance for SMEs;
- SMEs are active enough in utilisating results of the researches of Defence branches of Industry;
- considerable support of SMEs on regional level (in comparison with earlier existing situation) is principally a new situation in Russia.

"ECONOMIC REFORM IN RUSSIA AND GOVERNMENT REGULATIONS OF SCIENCE, TECHNOLOGY AND ACTIVITIES OF R&D SMES

FEDOROVICH V.A.
Head of the Division of General Problems of American Economy, Institute of the US and Canada Studies of the Russian Academy of Sciences
Moscow, Russia

1. ABSTRACT OF REPORT AT THE WORKSHOP

Now the governments of the United States and of a number of industrially developed countries of the West show their willigness to provide Russia rather with technical assistance than with economic aid. That means the assistance in building up a real market institutions and management mechanisms in science, technology and economy. This assistance includes also such field as mechanisms for state regulations in science, technology and economy. The strategic question is to substitute centralized planning system with such instruments as Federal Contract, state market for goods and servicesincluding R&D in particular, to apply principles of establishing competitive state market of goods and services and principles of state entrepreneurship's activities.
This approach of the West is extremely important for the process of reforming our S&T and industrial potential. The basis for it lies in real economic, S&T and industrial policy of the West. We, the represemtatives of business and science of Russia welcome such approach and we are waiting for such cooperation.
In my report I would like to treat the following issues:

1. Economic reforms and necessity to establish real market mechanisms of state entrepreneurship and state purchase of goods, services, R&D for the needs of national economic development and the state.
2. State market of goods, services and R&D in Russia problems of creation and the experience of Western corporations-contractors of the State for R&D Programmes.
3. State market of goods, services and R&D in Russia : the role of small innovation enterpises.

C. Corsi (ed.), Science and Innovation as Strategic Tools for Industrial and Economic Growth, 57-60.

2.

The year 1994 is the most significant frontier in Russian politics, in radical reforming of its social and political organization, in ways, methods and forms of carrying out economic reforms.

To overcome the deep economic fall down Russia need to apply new approaches, forms and tools in the process of reforming that really might correspond to the conditions and traditions of Russia in this period. What is the most important thing is that they should imply a commonly acknowledged experience of world entrepreneurship and organization of management for market economy of industrially developed countries of the West at the end of the 20th century.

In his address to the Federal Assembly of Russia the Russian President B.Eltsin gave road to new ways of overcoming the economic crisis on the basis of inroduction of state regulation of economy, determination of priorities in economic policy of state, creation of mechanisms of state economy. The address opens the door for critical evaluation and application of commonly approved methods of management andeconomy, escpecially of the state, of the Western countries.

The successive realization of the second stage of economic reform in Russia now needs to resolve a complex of social, economic and organizational and managerial tasks. They include:

1) successive privatization of the state property, otherwise formation of the material ground for the socially oriented "mixed" economy,
2) substitution of the former methods of planning the economy with the market type tools appoved in the world practice,
3) formation of the competitive market of goods and services,
4) formation of the all-national market of goods and services,
5) coming away from the principals of indirect regulation in economy and active transition to the state regulation in economy,
6) determination of limits of the state sector and of the main priorities of economic policy.

The world economic experience tells that instead of "the state order" (lever for the centralized planning, instrument of irretrievable loans and of legal robbing the treasury) we need to transit succesively, step by step to the Federal Contract System. In other words to push out the "state order" with the "federal contract" - the tools adequate to the market economy.

3.

We know that the world economic experience and state regulation in science, technology and economy have today as their base the two most important elements of market economy: state and all-national markets of goods and services. The former is the inalienable part of the latter.

The world experience shows that the state market in its modern form was formed in the second half of the 20th century, when in conditions of modern S&T revolution the state has been turned into the main entrepreneur and banker - the customer of goods and services, especially of the most new technique and technology. Finally the state sector of the Western countries (independently from property forms) includes usually key or strategic sectors of economy. They are S&T, fuel and energy complexes, transport, communication, defense, public health service, space and World Ocean. As you well know, the state uses for S&T development a comprehensivemechanism of state regulation.

The capital investments to science remain to be the priority function of the state. So, the state is in charge of the development of the areas that are economically or technically risky or could not be realized in private sector.

We firmly believe that now for the S&T and the economy
under reforms we need just the similar principals of management and the similar economic and technical policy.

This is of tremendous importance for the area of R&D that in all countries, one way or another, is supported by the state.

Small innovative enterprises - the engine of S&T progress - have special support from the state. Small enterprises are the basis and the most important element of all market economy and entrepreneurship.

4.

Deepening reforms, transition to their second stage require to resolve the four mutually connected tasks:

- successive privatization,
- successive formation of the state competitive market of goods and services,
- creation of a new economic mechanism - Federal Contract system,
- creation in line with market principles of the state S&T complex on the basis of the structures engaged in S&T that are being reformed.

The first steps have been made: there have been adopted the Governmental decisions on the successive transition to the "contract mechanism", contract economy. This contract mechanism should successivly become the cell of the private sector. The small R&D business has not yet become the most important legitimate contractor of the state. His role is interpreted in different ways by various political forces and groups of capital. In a number of cases large industrial capital tries to put apart the small business (in particular innovative) from the process of creation of new technique and technology.

5. CONCLUSIONS AND RECOMMENDATIONS.

Today when the administrative mechanism of planning economy in Russia has been broken, in order to substitute the overall centralized planning there should be created the state market of goods and services, the state order should be substituted by the Federal contract of Russia.

The necessary measures on the creation of the state market of goods and services should include the following:

- adoption of the Law on development of strategic areas of economy oriented to the specified programmes, the basis for distribution of federal contracts on deliveries of goods and services for the state consumption purposes;
- formation in the process of privatization of the most important element of the state market of goods and services the research industrial and technological complex of Russia (Research institutes, design bureaus, industrial enterprises, technological centres and laboratories, SMEs - the state contractors);
- giving on the mutual basis the right to the foreign capital to participate in deliveries of goods and services to the state market of Russia in condition when all forms of its economic activities would be guaranteed.

Creation of the Federal Contract system and the state market of goods and services in Russia will lead to the rationalization of the state economy, optimization of the budget expenditures : the principle of market economy "budget money should be given for the best goods at the minimal price" would be really implemented.

Naturally, these measures and transition to the contract economy should be effected succesively. Apparently, the process should be started in 2-3 leading ministries and institutions (Ministries ofsciences, economy, defense, industries, agriculture).

For these aims on a competitive basis there should be selected a number of large long term S&T and economic programmes for 5-10 years.

BRINGING RESEARCH AND INNOVATION TO THE MARKETS

UMBERTO DEL CANUTO
President IRI Management
Roma, Italy

1. INTRODUCTION.

As an introduction to the topic of Science and Technology Parks, which I will confront to hereafter, I would like to comment on the undergoing processes of trade-off between research on one side and economic development on the other.
We can look at this processes, of course, from the point of view of the industrial countries or from that of industrializing ones: while confining me to the first one, it might only be said that both process are increasingly approaching to one another under the impact of emerging technologies. In fact technology can be assumed, in particular as far as production processes are concerned, as the main integrator of different cultures operating worldwide.

2. RESEARCH, INNOVATION, AND ECONOMIC DEVELOPMENT IN INDUSTRIAL ECONOMIES: RECENT TRENDS.

I shall start considering the system of already industrialized countries, which suffer presently from adjustment processes due to changing technologies and from powerful switches in demand of consumer as well as of investment goods, led by the increase in per capita incomes.
This adjustments caused all industrialized economies to confront with sectoral crises and industrial reconversion: we will see in more detail the existing relationships between research, technology and overall innovation on one side and movements in the structure and organization of the production process (in a broad sense) on the other.

2.1 Industrial restructuring under the impact of technologies.

It is worthwile reminding us about technology paradygms: particular "filières" of products, moving from a unique set of scientific knowledge have impressively influenced, in different stages of economic growth, the economic environment and the production processes by which industrial goods, or services, were made available to consumers or firms.

C. Corsi (ed.), Science and Innovation as Strategic Tools for Industrial and Economic Growth, 61-70.

Railways and road transports, electrical power production and networking, chemical knowledge and new materials, more recently electronics and information technologies, have brought about incredible mutations on the production and social environment of mankind.
Technology changes have caused the rise and fall of whole industrial sectors, obliging workers to reconvert their know-how, to restructure education processes, to reinvent their lives.
Outcomes of this reconversion were an increase in social productivity, a decrease in hours worked, impressive take-offs of per capita incomes and - as a consequence of that a widespread structural adjustment in the internal demand: with an increase in services vis à vis of industrial and agricultural goods.
This bi-polar action, due to the technical change and to the structure of demand, is well known by the innovation scientists and labeled as a "technology push" phenomenon, as opposed to "demand pull" driving forces.
I need not go into details of a well known topic in innovation theory. Nor will I spend many words in reminding us all that innovation is not only a "product" one (although research is of paramount importance in this area), while "process" innovation gives also a major contribution in increasing productivity of firms and countries.
More will be said, instead, about innovations in productive environment: both on the side of externatilities and as driving forces organizing production through area networking, as opposed to past organization in large manufacturing sites.
Nations are increasingly in competition: efficiency inside a nation is the result of infrastructure investment largely financed and ruled by public intervention: either financial or entrepreneurial, or both.
State budget spending, mixed economy, nationalized industries, former central economic planning, are all sides of one only coin. State intervention alone can (rapidly) overcome lags in infrastructuring and enhance productivity and competition of industrial countries: a necessary condition, even though not a sufficient one.
The result of state intervention is a broad networking: power, oil and gas, telecommunications, roads, railways, and again banks, insurances, education. Networking environment has a deep impact over labour organization. It changes the focus from the Taylor assembly line to organization over territory: large enterprises and their subcontractors (virtual corporations), industrial districts.

2.2 Services diffusion in human society, scientific research and teaching.

Employment in the services sectors exceed 60% in industrialized countries. Only five decades ago Italy had about 30 people over 100 employed in agriculture.
We still perceive industrialized countries based on manufacturing: the conceptual framework we use - factory production, productivity increases, blue collar work, cost push inflation, product or process innovation - is heavily dependent on "industrial" rather than "services" thinking.
Yet it is no longer so; or, at least, the culture of quality, typical of a "service", has deeply influenced manufacturing. Customers' fidelization is influenced by after

purchase guarantees, quality assurance, safety controls, environmental responsibility: industrial products are conceived differently than two decades ago, with a larger content of "services" in and around them.
Markets are worldwide: goods' availability depends on transportation services, global marketing depends on telecommunications, finance and insurances are international.
The "services" culture influences strategic organization and thinking: life-cycles are shorter, due to imitation, simultaneous marketing leads to joint venturing, strategic decision making is peripherized rather than based on central planning.
A third dimension is added by information technology as a paradygm. While other techniques (power distribution, for instance) require a passive attitude from the consumer, both hardware and software devices depend on an active role of the user, to be adapted to his needs through skills acquisition and organization development.
As a consequence of that, education and training processes (again services) acquire a major role in production enhancement. Diffusion in education is a generalized feature to achieve success in worldwide industrial competition. That adds up to the need of high education standards to cope with science problems. University teaching is no longer sufficient, and must be complemented with post secondary mass education and continuous post graduate adjournment.

2.3 The sunset of the "large corporation" model.

For the last two decades the "large corporation" model has been fading out. Operations and sales have been externalized to subcontractors; central functions remained marketing, finance, quality control, core technologies, networking, strategic management.
A central "lean corporation" generated, through networking, a larger "virtual corporation " including subcontractors, tied by market links and by technology identities.
Research split over the two: only core technology remained proprietary, while subcontractors were committed in subsystem innovation and related research. In the small and medium sized enterprises milieu a different model emerged: the industrial district.
Industrial district can be defined as a form of spontaneous organization of the coordinating functions of the large corporations, occurring in a local environment where several SMEs operate in a narrow range of products (or services). It is a concept that goes back to the last century, when it was theorized by Alfred Marshall.
The basis of the self-organization mode is usually found in the fact that several firms work in the same "milieu": the contiguity assured by this fact has a strong influence on the product: actually only a narrow range of similar products is usually found in the same industrial district.
The "milieu" assures process technology, marketing outlets, core "product technology" (including time to market and product-cycle when products are changing); on the other hand logistics, financial services, specialized workforce, image, quality are organization problems, common to SMEs operating in the same district, and are jointly confronted to

by cooperators-competitors of the same area, with the help of local authorities.

3. CHANGING THE STAGE AND THE PLAY: OLD ACTORS FOR NEW ROLES AND VALUES.

Current intervention in research and technology is broadly confined to the States, to strong intermediate subjects (as Research Councils, Universities, Science and Technology Parks) and to firms (fundamentally to large global corporations, but also to the small and medium sized enterprises in industrial districts).
The three groups are the actors operating on the R&D stage.
Will - in a new global environment like the one we tried to depict in the previous part of this paper - their present roles change, or (in a different way of asking the same question), will their value systems change?
Trying to give a very brief answer to this actually complicate and cumbersome question, we will note three directions of change.
The States will move from the present intervention, based primarily on financial support, to a more active role of strategic guidance: R&D will become more and more an externality to the competitive system of the country, and strategic top down approaches will be necessary for an overall orientation of all the actors on the stage.
As far intermediate entities (as previously defined) are concerned, new goals are to be added, in perspective, to the old ones: Research Councils should give more attention to innovation, beyond research; Universities will be confronted with teaching of large numbers; Science Parks should become a powerful multiplier of employment.
In general, all the "excellence" orientation will be obliged to give more attention to "diffusion" phenomena.
The strenght of a nation will be based more and more on culture, diffused technology, brain intensive production, if the nation itself is suppose to keep competitive.
In the system of firms, innovation should be the focussing point: bringing technology to demand, coping with external competition, cooperating with State and Research Entities in pre-competitive technology.
These changes can fairly be accepted on the ground of principles directing the common action world over.
But the problem, in practice, is much more difficult: what will involve, in each culture, the practical realization of needed actions? How will react Europe vis à vis of Japan or United States? European culture does behave coherently, or national peculiarities are to be taken into account? Cohesion values are to be taken as top priorities, or local cultures are to be kept alive?
We shall try to answer those questions moving a step backward to consider how the different cultures have been defining the behavioural patterns of competition and of Science and Technological organization in the very last decades.

3.1 Public intervention and its strategies in research related job creation.

3.1.1 The Anglosaxon world: two models.

The history of growth of the Science Parks model shows important differences around the world, due to local approaches and to different cultures or government values.
The originary model of a Science Park goes back to the fifties and to the United States: top level Universities (Stanford and MIT) decided to cooperate with top-level corporations operating in electronics at very high levels of technology (IBM, AT&T) .
Necessary proximity of research centres was the first core of what afterwards was named Science Park (Silicon Valley, Route 181); the content of it was only a close collaboration on the border of the science paradygms between top level Universities and top level large corporations.
The idea of moving this model to other Universities in the anglosaxon world proved in general to be unfruitful: top level Universities are actually scarce resources and joint venturing with number one corporations in technology was obviously confined to a few occurrencies.
In fact the tentative to involve Universities in Science Parks shaped a new model in England and the United States, where Universities operate with large authonomy and own large estates. There was wide room to cooperate with small and medium sized enterprises, growing incubators to offer them scientific and technological, as well as business and entrepreneurial, support.
This model, based on adding value to physical infrastructures owned by the Universities, was coherent with their needs to enlarge their authonomy from State resources, due to the decreasing aids from the Government budgets; the result was a large liaison with industries to which Universities offered consultancy, technology transfer and incubators.
Very frequently teachers themselves moved into spin-offs: when technology and business sensitivity met, important breakthroughts emerged (Apple, Genentech); in ordinary cases technology diffusion was a primary outcome of this University-Industry cooperation in the anglosaxon world.

3.1.2 The european experience.

The anglo-saxon model, essentially based on Universities estates value-adding, and therefore on a bottom-up approach, was taken in France by the government and ruled in a top-down way.
French government involved two Ministries - the Research Ministry and the Environment one - in networking at local level. Two organizations operate: in incubators (DATAR controlled by the Ministry for Environment) and in financing and disseminating research through technology transfer (ANVAR, depending from the Minstry for Research).
The outcome of this policy was the creation of a large research oriented "sunbelt" in the French "Midi", rich with Universities, Science and Technology Parks, research centres (CNRS) and large advanced technology corporations (IBM, Aèrospatiale, Thomson). "Brain" was moved from "central" Paris to the South, conditions were created to

"diffuse" technological selfconsciousness and market oriented mind.
A different approach was taken in the german world and in the northen european countries. It is well known that technology and teaching are committed to german "States" (Laender); localism is also very strong in the other northern countries.
Local approach proved then to be fruitful to innovate and to commit University education to helping growing industrial sectors: giving advise in product and in process innovation as well, and linking local management to the overall european world and to cohesion aids of the European Union.

3.1.3 Japan.
Last but not least, as an example of a different cultural approach to the problem of technology transfer, the japanese mode is worthwile being quoted.
All of us know about the MITI (University of Industry and Foreign Trade) coordination role of large japanese groups in long term (ten years) programmes. We know about science cities like Tsukuba; we know about peripheral experiences of Business Innovation Centres in the less favoured areas of Japan.
I want only to note that cooperative action adds up financial resources, giving synergies to amounts of money devoted to research, sometimes of a lesser extent than in other countries.
On top of that, agreements are taken by the participants to exploit the common know-how, orienting it to the strategies followed on the international markets by the "Japan multinational Inc.".

3.1.4 The italian situation.
This being the overall worldwide panorama, which is the Italian situation?
Other papers will give important details on the experiences of my country, in this same conference.
I would only like to note that the vast phenomenology of italian experience can be standardized in about five broad groups.

1. The first one concerns Universities, their managerial authonomy and the creation, aside them, of industrial liaison offices (generally in the form of Consortia), including large companies, Research Centres, Chambers of Commerce, Local Authorities and other entities. This experience can be named the experience of City Research Consortia, promoted by my company, IRI, in the mid-eighties.
2. A second group of experiences is to be included in the standard model of the Science or Technology Park: large estates with science laboratories, incubators, technical opportunities, international liaisons: just to name some of them, this is the experience of Bari's Technopolis, of Trieste Science Park, of Brindisi's CNRSM in new materials.
3. Then we have networks of BICs: the most important on is that of IRI's SPI: it includes many incubators with a precise commitment in job and enterprise creation in a regional framework.
4. Following the same pattern, but specialized in intervention to develop italian southern regions, is of growing importance the experience of southern Scientific

Parks, financed and vetted by the italian Ministry for University and Scientific Research.

5. Last but not least, a new model is appearing to the stage, in western Tuscany: a Regional Science Park, joining the forces of all the operators in the region (at scientific as well as at financial or local authority level) to give a more precise coordination and orientation towards a restructuring of the regional economy, based largely on the presence of interesting industrial districts.

4. SCIENCE PARKS AND TECHNOLOGY TRANSFER: CURRENT MODELS OF DIFFUSION AND THE ROLE OF INTERMEDIATE LARGE ENTITIES. FUTURE PROBLEMS AND TRENDS: FROM A TECHNOLOGY PUSH TO A DEMAND PULL ORIENTATION.

To end this very brief comment on undergoing trends in technology transfer, I would like to point out some open issues and the problems related to them.

The general trend underlying the factual phenomenology - I suppose worldwide, but certainly in Italy - is to move the model of State and related intervention in research and innovation from a technology push approach to a demand pull one.

Researching, in a country, is an activity deeply interrelated to other public goals.

Education, is the first, both as a creation of excellent brain resources (which is a top level Universities' job) and as a diffusion process of knowledge, through compulsory school and education patterns oriented to the use of "brains" inside the firms.

A second goal is job creation: competition worldwide is more and more linked with self-empowerment, with open organizations: continuous education is a "must" for firms, average levels of training must be raised.

School and research are a necessary seed for production and welfare. They are instruments for enlarged goals, pertaining to the "welfare of nations".

4.1 The international control of technology "filières".

Technology has a globalized, worldwide approach.

Some countries have a leading role in developing new "filières". Other countries are followers. There is a "make versus buy" approach: the premium price of "makers" is something eroded by fast followers. In any case technologies must be "applied" in production: which is always a "make" (organization) process tied to local values and based on "processes" bought and adapted to local strenghts and weaknesses.

Which is the role of Universities in this framework: researching, high level teaching, transferring technologies, education diffusion?

In parallel to this, which is the mission of large Research Public Entities, financed and ruled by the States?

Some precise trends are emerging: without inferring anything, we shall note them.

There is a wide lag between the technology border and the potential capability of market innovation. The lag between science border and market is even wider. We must fill those lags. We need linking instruments: High Research and Education Institutions

are the first source to fill this gap.
My suggestion is not perhaps in line with the "pure science" supporters, but it is obvious that this role of "international gatekeeping", vis à vis of emergent worldwide technologies, must be organized and linked with the production world of large enterprises and SMEs.
This integration is of primary importance at international level: Universities and Hight Research Institutions have, in any country, this central role: not (only) to be researchers, but to be gatekeepers.

4.2 Technologies and markets: shared know-how and diffusion models.

We have already noted that technology and science are a long way ahead of the markets.
The problem of technology transfer organizing is essentially a marketing one and falls inside the goals of the strategic organization of the production system, as oriented to what we call innovative process: a goal which is essentially a "demand pull" one, led by local cultures.
How organizing market oriented innovation?
As an example, I shall refer to three important models that can be considered as integrators between markets and shared know-how.

1. The first model falls inside the realm of what we called "virtual corporation": the complex relationship joining large firms (also in the services sector) and their "subcontractors".
   Technology and training (know-how) have, so far, been considered as "proprietary" to the large corporations.
   When those entities extend to their subcontractors, also technology transfer (in both senses, also from subcontractors to the central core) and related education and training initiatives must be conceived as a problem of networking, whose caracter is to be "open" to the outside world rather than restricted and owned in secrecy by the firm.
   Organizing innovative links in this networking is perhaps the most powerful device to create and diffuse innovation.
2. A second approach is to create - either in a bottom-up way, like in the italian case of industrial districts, or in a top-down one, like in the french experience of ANVAR - appropriate structures oriented to ease the technology transfer.
   This is a "supply" approach, and therefore is a weak one, as it does not organize the market demand. Commonly tech-transfer agencies can give answers to firms: but SMEs very seldom "demand". This is one of the reasons why italian industrial districts, with their contiguity and the common "milieu", perform better than agencies.
3. A third approach is to create commercialization structures that can give advice on intenational markets and on necessary product modifying to enter them.
   The most performant model is the one of japanese "sogo-shosha" (trading companies): it is very difficult to move their experience in a western "competitive"

and WASP culture, but nevertheless I do not see any other alternative approach to organizing market penetration in a regional, national or supra-national way.
All the three examples referred to above have one issue in common: technology transfer is a marketing problem, not a technology one. No solution can start from University or from supply: demand must be organized in a cooperative, not in a competitive way.

4.3 Large corporations and joint research programmes: cooperate or appropriate.

A last notation should be devoted to large corporation, and to their own core technology research.
Large corporations in each country and much more in Unions, like the European one, have a major role in leading the technology paradygms which are particular to the "filières" in which the country or Union is leader.
Leadership in the chemistry in the twenties was a german one; at the end of last country electricity and telephones developed mainly in the United States and in certain European countries; telecommunications nowadays have major strongholds in the United States and Japan, with important contributions of Europeans; consumer durables and electronic appliances have focussing points in Japan, while computer science and software share equally their development between U.S. and Japan.
Beyond any example, the technology paradygm is led by certain Company Groups the world over, supported by definite countries or Unions. This leadership must be preserved, and the common effort is one of the major tasks of the large Groups of firms which we usually call "multinational corporations".
States and Unions use to provide them with financial resources; sometimes, as in Japan, we see that the State itself gives a strategically major contribution also in orienting and monitoring the research, through decadal programmes.
This is the "real" process which is usually hidden behind the "struggle for figures" about financial aid to research, given by single OECD countries.
A "real" process which is based on different models: public spending and "laisser faire" typical of U.S.; major commitment of the Government and Keiretzus in Japan; Framework Research Programmes in European Union, "Eureka" market oriented research in enlarged european environments.
What is the core issue in all these cases?
Apparently there is an hidden conflict between "pure" and "applied" research, between "pre-competitive investigation" and core products "applications", between secrecy and patents.
This contradiction has to be solved, hopefully in cooperative ways rather than in conflicting rows. It is evident, for instance, that sometimes the patent system hinders technology transfer rather than favouring it.
Time to market and shorter product life cycles are the core strenghts of the firms, in parallel with worldwide organization, to create value and to generate profits. Competition is more and more a marketing than a research and technology problem.
We should cooperate in research and compete in marketing.

We should organize, also at the level of competing multinationals and countries, cooperation in research and technology, even in different groups of countries around the world: cooperation should be beneficial to all.
Whereas competition should be confined in the world of marketing, in a "demand pull" innovation, where products and their quality should be the focussing point to orient the customers' choices.

INNOVATION IN CRISIS: HUNGARY BEFORE AND AFTER THE WATERSHED OF 1989

The paper is published in **"Technovation"**, *Elsevier Science Ltd., Oxford, England, Vol.14, No.9 (1994), pp.601-611*

BORISZ SZÁNTÓ
Academic Research Unit, Technical University
Müegyetem rkp 9, T ép, 201/b
H-1111 Budapest, Hungary

1. ABSTRACT

The same process of globalisation of the technology, which causes the economic failure and disintegration of Eastern Europe forces Europe to integrate. The aim of this paper is to show, that Eastern Europe has found itself in the process of regression with its ability to innovate declining. The analysis of conceptual hindrances refers to the crisis of the system of values, while the analysis of parameters signals the crisis of innovation. The change for active goal-setting and problem-solving activity on the governmental level is urged. The integrating Europe should actively promote in its Eastern part the application of the paradigm of technology policy.

2. THE PARADIGM OF TECHNOLOGICAL CHANGE

Technology policy has been portrayed as a techno-economic paradigm of industrially rapidly developing states and as an answer to a new phase of global development called the era of globalizing technology. This new pattern of management was an expression of Schumpeterian innovative entrepreneurship at the governmental level. Technology policy is the policy of economic success through implementation of competitively created technological advantage, the policy of "changes in economic life as are not forced upon it from without but arise by its own initiative from within" (Schumpeter, 1961, p.63). Compared to the paradigm of "mere management" of passive adaptation enough to sustain economic equilibrium and stability, the paradigm of innovative entrepreneurship involves efforts of will to conceive and carry out projects of competitive technological advancement and its economic realization on the world

C. Corsi (ed.), Science and Innovation as Strategic Tools for Industrial and Economic Growth, 71-83.

market. Both paradigms, the paradigm of routine economy striving for stability and the paradigm of technological change causing disequilibrium, despite of their obvious conflicts can be maintained collaterally due to application of specific social and economic mechanisms. This coexistence of contraversial paradigms is the most striking proof of a higher level of socio-technical evolution the highly industrialized countries have arrived at.

Hungary has arrived at the drastic political changes of 1989 without clearly formulated and properly functioning innovation and technology policy. It is true, that the importance of innovation and technical development has been stressed during the last decade in innumerable governmental declarations and decisions. But in reality innovation and technology policy has never been - with a few exceptions - and still is not considered the pivot of the governmental economic policy. Even in 1993 the formulation of principles of innovation strategy was considered by Governing Principles of Science Policy (TPB, 1993) as a task of secondary importance. This document preferred to dive first priority to the "optimisation of science, education and R&D", together with "framing of healthy historical consciousness of Hungarian nationhood".

Science policy, as the policy in science (and not the policy through science), the policy of scientific establishment, has always predominated in Hungary. Nevertheless, in May of 1993 the concept of Governmental Innovation Policy was formulated and finally accepted by the government. This truly professional document (OMFB-IKM-PM, 1993) is at last in full correspondence with related principles of technology policy of leading OECD countries. It is clearly stated in the concept, that even well developed and dominant free market forces are not to foster innovation without proper governmental orientation.

The present study and the analysis of conceptual hindrances are based first of all on my own theory of socio-technical evolution (Szántó, 1990). The system of values I am taking as guiding light in the still prevailing East-Central European kingdom of old values is that of the technology policy of the leading countries and the recent concept of the Hungarian governmental innovation policy.

3. THE SYSTEM OF VALUES AND THE BURDEN OF CONCEPTUAL INHERITANCE

In principle, technical development does not depend on the political structure of the country and the official ideology. The solution of technical problems requires pragmatism, knowledge and imagination. But technical development does depend on general and technical culture, traditions, practice and acquired experience, level of development, and of course on the strengths of economy, market demand, degree of entrepreneurial autonomy, and the ethical environment, the functioning system of values of the society.

The system of values that regards technical development as a matter of mere investment, with emphasis on production, is the heritage of the heaviest burden in East-Central Europe. Investment was and still is considered here mostly the only way of technical development. Its task was the maintenance and enlargement of production, and reproduction of the system of values. This theory has been in absolute accordance not only with the concept of central planning, but also with the leading conventional Western and Eastern economic theories. According to this philosophy, output is the function of input. In other words, if the state has invested into R&D, it is entitled to expect innovation and profit at the end of the tunnel. The centralistic hierarchy of the party-state was the best structure for concentrating centrally planned investments and selecting segments for development.

The concept of innovation policy, on the contrary, states that the accent of economic policy should be placed on the development of individual innovation capability. Investment is nothing else, then rendering to the well functioning innovative entrepreneurship the means of turning technological advantage into market advantage and, later on, into routine production and economic benefit. First of all, it is innovative entrepreneurs' individual problem-setting and problem-solving that is to be supported by the state, banks and specific socio-economic mechanisms.

Huge industrial monopolies were formed in former socialist countries as a means of concentration of capital and labor. Austerity measures, full utilization of industrial capacities, and keeping clear off the unjustified parallelism in industry were to serve the goal of concentration of the capital. In principle, these measures were also justified by mere economic logic. But, as the result of this investment policy, the great diversity in the economy, required as a precondition for the rather improbable emerging of a few innovative entrepreneurs, has gradually become emaciated to the minimum.

The ideology of the "follow-up development", levelling, bringing up the backward by centrally planned investments, had another equalising effect in economy: there was no intention of considering individual innovative ability as a prime factor of economic progress. This was a hindrance indeed for those few managers of state companies who happened to be more talented and ambitious than the average during the last 40 years. Such managers could probably succeed in some areas of economy, but their achievements usually proved to be temporary since their applications for further state capital investments for their second and third innovative stages were turned down on the grounds of the concept of equitable distribution.

Modern innovation strategy, on the contrary, relies on and supports innovative entrepreneurs who manage to be ahead of others and show the ability to develop themselves to even a higher technological level, thus pulling up the rest of the industry. The strategy of innovative development is required just for the very reason: to form the spearhead of technological progress and to shift on the top of their technological level from one technology to another, even more promising and sophisticated one, keeping the pace of development and dynamic equilibrium of innovative change.

The Soviet model of central planning and control divided the process of innovation into separated stages, linearizing the nonlinear. Isolated factories and institutions were not to collaborate with each other without special permission from their authorities. According to this philosophy, the process of innovation is a sequence of fragmented routine activities, a chain of phases from research to sales, with production playing the crucial leading role. This theory has been also in absolute accordance with conventional economic theories. As a result, continuous feedback and close cooperation between

potential partners and would-be system components has almost ceased to exist, their natural functioning made almost impossible. The innovative ability of the industry gradually withered away. Commands and compulsory planning were to replace the essentially individual problem-setting and informal cooperation. The role of the individual was rejected on ideological grounds.
The modern innovation theory regards innovation a nonlinear teleological social phenomenon based on individual ambition and willpower, with a strong intention of individuals to cooperate and integrate their efforts.
The inventory of conceptual hindrances can be continued. To mention just one more: innovation as socio-economic phenomenon can neither be expressed nor explained by the theory of labor value. However, all the financial regulations of technical progress were, and still are, based on this concept of conventional economics designed for routine activities.
The conceptual inheritance would not even be mentioned here were most of the misguiding principles not still the part of the contemporary system of values in the East-Central Europe. As is known from the work of Mihály Polányi, the system of values lives in everyday practice and is not expressed openly, yet is the embodiment of dominant philosophical concepts (Polányi, 1992 p.179). And this is the present day philosophy of economic policy of Hungary and Eastern Europe. It has given up central planning but not the system of its values. Creation of a new ideology, or returning to the values of the past, obviously does not breath life immediately into a new system of values of society even if the old one has been proclaimed to be abandoned together with the political structure. The theory of socio-technical evolution says that no one can function without a model in his mind, and it is the whole of the hierarchy of the society that influences each entrepreneur as its part through the dominant system of values, placing before him the paradigms and models it prefers. The Schumpeterian paradigm of risk-taking innovative entrepreneurship on both the personal and the governmental level is obviously not the model of the dominant system of values these days.
The reigning system of values can be easily spotted, for example, by the mode and principles of redistribution of financial resources and the state-given privileges, or by the specific characteristics the sensors of control are tuned to 1. Innovation and technological advantage were not the objects of preference of the system of redistribution of resources in Hungary either before nor after 1989. What is considered important is mere survival. This fact shows that the system of values in former socialist countries is still in crisis, as is the whole economy of the region.
One can notice this from the fact that the horizon of economic and social decisions keeps getting shorter, both at company and at governmental level. People are apt to choose models for functioning from their 40-years-old past. Long-range visions and development projects are usually hardly comprehended, governmental preferences are granted to everyday cash inflow, sensing receptors of control are fixed on the problems of today - these are the signs of the crisis of the system of values. The climate, the system of values, should rather make entrepreneur and supporting institutions regard market failure less risky than the atrophy of innovation initiative and ambition. Shortsighted monetary policy and fire-fighting in economy signal a system of values

hostile toward innovation. Officially, the philosophy of economic policy of the centralistic hierarchy was abandoned on the watershed of 1989. In reality its heritage is still with us, blocking the way of socio-technical evolution.

4. THE BURDEN OF STRUCTURAL INHERITANCE AND THE EARLY REFORMS

Structure and its rigidity are very much connected to the system of values and the philosophy of economic development. Structure also means an inherited paradigm, a habit and a certain attachment to inherited models. Restructuring is mostly blocked by people's clinging to customary things.

The structure of industrial companies still bears the marks of the centralistic hierarchy and its logic of commands from the top. The planning economy got rid of the plurality of small and medium size supplier companies around bigger ones, almost eliminated the cooperative culture of suppliers, and left behind a number of vertical monopolies. In spite of rapidly growing number of private companies since the early 1980s and especially 1990, industrial structure shows only a few signs of recovery.

Foreign trade companies, created by the logic of concentration, structurally detached marketing from R&D and other activities of innovation. This prevented technical development from reacting to direct feedback from the market, even in the 1980s when licences for independent foreign trade were granted to industrial companies much more easily. There was a chance and a definite intention in the middle of 1980s to transform the structure of foreign trade companies into innovation-oriented market- and industry-organizing financial network, something like the Japanese ***sogo sosha*-s** (trading houses). This opportunity has been lost.

The party-state reserved to itself the task of labour division. As a kind of Maecenas, the patron of sciences, it separated world-famous research laboratories from the industry and locked them into a privileged reservation of research institutes. Those institutes provided relatively good conditions for scientific work for quite a number of intellectuals. After 1989, the economic crisis made even this heritage, positive in its cultural effect, a burden which enfeebled the state budget. On the other hand, the role of engines of innovation assigned to scientific institutions is still a major hindrance for the inherited science policy. Because of this, science policy in Hungary looked to innovation through the eyes of the research establishment and was never able to deal with innovation with a proper degree of importance. It can be observed that science policy and technical development policy were, and still are, two inherited conflicting paradigms, due to their diametrically opposite orientations (one toward the stability of undisturbed self-oriented research, and the other toward the change of production) as well as the "either-or" resource allocation policy of the government's budget.

Paradoxically, the early reforms of the 1960s placed the policy of technical development under the necessity of more intensive defence. The concept of free market forces started gradually to grow intolerant and bigoted toward the concept of "centralistic" technical policy, which had been left out from the main stream of the reform of 1968. Even in 1993, the inherited structure of governmental bodies makes innovation and technological progress the matter of separate responsibility of the State Committee of Technical Development (TPB, 1993), whereas it is the Ministry of Finance, the Hungarian National Bank, and the Ministry of International Economic Relations that are really influencing the companies in a much greater degree. The Ministry of Industry and Trade also bears responsibility, but has almost no means of influence.

There is always a contradiction between short-sighted market or monetary policy and a long-range technology policy. In any country, innovation is always a disturbing factor for routine production. In highly developed countries this natural contradiction is solved by a number of special socio-economic mechanisms, which allow both paradigms to function in the evolving society. Hungary was the only country in Eastern Europe where the crucial role of innovation was recognized in the late 1970s and some efforts were made to increase innovative activity. The "anti-centralistic" conceptual bigotry and the increasingly monetary character of financial policy mostly neutralized even those few mechanisms that where designed and introduced in the 1980s to foster innovation in Hungary. The Innovation Fund was created, for example, in 1980 as an autonomous bank with the task to assist Hungarian inventors. Over a period of one year four other banks were created by involved ministries. Those organizations were far from real banks, but they initiated a process of disintegration of the monolithic banking system. It is very characteristic of the reaction of financial policy makers, that in 1986 the State Planning Office elaborated the project of a Central Innovation Bank by integrating all the small banks into one and therefore terminating their independence. In 1987 all these banks became commercial banks, restructuring the monolithic banking system into a two-level system of banks with growing degree of independence.

5. INNOVATION IN DECLINE

Declining innovation activity, in its relative terms, means growing backwardness on the world market. In Hungary innovation activity declined in absolute terms as well, not only becoming less and less significant in comparison with the highly industrialised countries, but year after year losing its strength in comparison with its own earlier performance. According to the theory of socio-technical evolution, this kind of evolution process can be considered as retrogression or regression. Regression is a phenomenon of the process of decay, characterized by weakening answers to the ever stronger challenges from the outside, and the progressively weakening ability to adapt. The theory says that there can be only two forms of evolutionary movement: progress or regression. Stagnation is only a tentative and short period of hardly changing performance of an evolving functional system.

Having got into regression, companies usually try to survive by sacrificing part of their property and resources, narrowing their field of activity. Each stage of the reproduction of the past brings for them worse and worse economic performance, uncertainty, loss of values and longing for stability.

Statistics can observe innovation only when it has become a routine, and not a curiosity, as it is today in Hungary. It is therefore rather difficult to demonstrate the process of decline on the basis of official statistics and observations, but there is no other choice.

Education in socialist countries has always been a matter of prime importance. In Hungary, quantity and quality parameters of education have been on the level of European standards during the last 40 years. Expenditures reached 5% of GDP, which corresponded to the ratio in highly developed countries in Europe. Technical education was traditionally good, despite of some concessions on the quality. The backwardness and time-lag happened because of shortage of founds, low wages, and the counter-selection this caused among teachers of lower grades. The result is that high-school attendance was 71% in 1988, while in OECD countries it was 95%, in Taiwan 90%, in South-Korea 87%. University attendance in 1988 was 15% only, while the ratio in the countries above was 41%, 37% and 32% respectively. Intellectual impoverishment is the greatest danger for a country, because it can stop its socio-technical evolution and push it to the periphery. This trend should be considered as very alarming.
The rate of diffusion of technology or the frequency of innovative entrepreneurship should be theoretically be taken as a very significant parameter of the country's innovation activity. Unfortunately, proper statistical data are not available even in industrially developed countries. In Hungarian industry, with its monopolistic structure, the rate of technology diffusion has always been considered very low. There is nothing encouraging for the time being either, since big state companies still form the larger part of the industry. The number of start-ups, growing rapidly, by tens of thousands especially after 1989, suggests that the number of degrees of freedom of innovative enterprises has also increased. On the other hand, experts estimate that the actual number of these entrerprises is much lower than before 1989. Some people regarded the activity of multinational companies as the only significant factor of technical development and diffusion of technology after 1989. But, up to 1993, their investments in Hungary did not exceed 15% of the total.
Original products and technologies did not play significant role in the country's economy. Within 50 years the proportion of the original, market-influencing products and technologies of the Hungarian industry fell from 30 per cent to 3 per cent (Szántó, 1990/5-6). In the last 15 years 40 per cent of the industrial products have not changed at all. The proportion of new products fell back from 2.5-3 per cent average between 1975 and 1980 to 1.8 per cent in 1983 (OMFB, 1986). Industrial products rapidly became outdated. The average age of Hungarian industrial products at the end of the 1980s was almost 16 years. The innovation period, from the idea till its market introduction as a product or technology, still takes in average 10-15 years.
The process taking place in the Hungarian agriculture in the late 1960s and 1970s has been quite different and has resulted in rapid innovative development, unique among the "socialist" countries. In 10 years, Hungarian agriculture has found itself among the leading countries with regard to productivity indexes of 42 products. The dynamics of development exceeded those of most European countries, while the cultivated area and the number of people working in agriculture decreased.
The paradigm of technical development in Hungarian agriculture between 1960 and 1980, with its innovative entrepreneurial values and system approach, was basically different from that of industry. The comparatively good results, however, were naturally accompanied by increasing investments. Thus the scarcity of resources and the crisis of

the whole economy stopped and even retarded the innovative development of agriculture in the second part of 1980s. The loss of the Eastern market, and the silly and incompetent agricultural policy at the beginning of 1990s, knocked out this once prospering economic sector. That, however, does not reduce the innovative model value of the Hungarian agriculture of that time.

Industry has basically become "reproductive" in its character during the past 40 years. It has been much more able to copy than achieve results independently. The productivity ratio did not reach 1%, and was 0.2% between 1976 and 1983. Industrial production dropped by at least 60% after 1989. but in the middle of 1993 the decrease seems to have stopped.

The investment ratio in industry was 50% in 1967, in the 1970s 7.7% annually, between 1975 and 1980 2.3% annually, between 1980 and 1987 1.9%. Between 1989 and 1992 the ratio decreased by 25% (on comparable prices), sinking below the amortization level. Import of machinery has fallen particularly drastically.

The age of equipment relates indirectly to the decline of innovative ability. Decreasing investment led to the fact that 40% of industrial machinery was over 10 years old in 1986, one fourth of which still operated when completely written off (OMFB, May 1988).

Regarding the level of development of industrial infrastructure, Hungary was 20th of 27 European countries at the end of 1980s.

Nation-wide R&D expenditures decreased from 2.32% of GDP in 1988 to 1.69% in 1990. R&D expenditures of industrial companies decreased in real terms in 1990-91 by 45%, their contracts with institutes and universities by 62%, their expenses on R&D equipment by 84%. Most of the research institutes, with few exceptions, were fighting for survival between 1990 and 1993. The cancellation of state R&D programmes and orders by 30% in 1992, the cut down of budget resources, the insolvency of the companies, the disappearing Eastern foreign market and shrinking domestic market, the abolition of all special favorable conditions for R&D activity, the drastic fall of contractual incomes compared to the extremely successful 1989, the dramatic reduction of industrial research activity, the decreasing staff2 and the low income of the employees, the lack of current funds for contractual activity (OECD, 1992), the fall of the government's Central Technical Development Found expenditure by 55% in 1992, the impossible situation of the partner institutions in the former socialist countries, the increasing rate (15-30% in 1991) of foreign working contracts of research fellows, the deteriorating material and financial conditions of the institutes and the problems of further financing, unemployment or "manpower rationalisation", competition, uncertainty - all these factors were very unfavorable for innovative activity in Hungary.

Licence import has been an insignificant factor of technical development in Hungary (OMFB, 1982). The knowledge injected into the country in the form of licences and know-how is very little. In 1980s the country spent hardly more than 0.1% of its GDP on licences and know-how (the Netherlands spent 4%), and spent 20-25% even less than that in 1991-1992. Licence import within the total technical development expenditure was 4.7% in 1969 and 9% in 1978, in absolute values the latter one was five times higher (OMFB, June 1982). Comparing this, though, with Austria, for example, where they spend more than 45% of the total R&D expenditure on technical development, not to mention Japan, we can conclude that it was rather low. The ratio of licence-based products in Hungary in 1989 remained less than 10% (OECD, 1993). Comparing licence export with import, we have to admit that the technological balance has had a negative tendency during the last 10 years (OECD, 1993, p.36), due mostly to the relatively activ development of agricultural and chemical companies (Table 1).

year	1981	1982	1983	1984	1985	1986	1987	1988	1989
M - HUF	435	45	368	46	95	458	506	996	1888

Tab. 1 Technological balance

The adaptation time for licensed, foreign technologies was 4-5 years, which does not cast a favorable light on our cultural infrastructure, the worker's ability for socio-technical adaptation and gradual change, considering the fact that the adaptation time in Japan is calculated in months.

Considering the actual number of domestic patents, Hungary was one of the least in Europe by the mid-1970s. By 1986-87 we had fought our way up to the middle of the league, and the number of domestic granted patents rose threefold. Thus property right regulation by registration of invention gained a larger importance for Hungarian inventors than that in the more developed countries, since there was no other possibility of personal reward and income. At that time invention was actually the only way of free and unlimited entrepreneurship in Hungary. The spread of the culture of industrial property rights was significant, and it met the standards of the leading countries in Europe.

According to the patent statistics, innovative ability was higher in some branches of industry than in others. Hungarian chemical industry showed higher R&D activity and owned half of patent applications registered in the USA (OECD, 1993)

years	1981	1983	1985	1987	1989
patents	98	107	108	126	131

Tab. 2 Hungarian chemical patents registered in the USA

The number of patents dropped fast in the early 1990s, bringing Hungary in this respect again to the low level of the 1970s.

Regarding innovative suggestions of company workers, however, we have been at the lowest point for quite a long time, with 0.01 suggestions per capita per year (OMFB, 1983). According to Swedish foreign service figures, this number was 16 times higher in the USA, and 2000 times higher in Japan in 1986. As a consequence of the introduction of "market conforming", but not really innovation-friendly taxation in 1988, this number even fell back by another 30%.

We must add that innovation in Hungary has always been largely dependent on the standards of business management. At most companies and research institutes the management culture, compared to the developed countries, is at a medium level. That means less ambitious managers who are ready for simpler forms of cooperation,

avoiding risk-taking, preferring mostly short tasks and striving for stability.
The concept and practice of contract-based research, as an additional source of income, has spread in the low income universities as a venture of the departments in the 1960s. This so-called "KK" (outside-budget income) spread parallel with the worsening conditions of state budget financing and loosening regulations. Gradually it has become the main source of income and research funding for several university departments and research institutes. In spite of the opposition, the process swept away in the 1980s all research (except part of the theoretical and basic research) oriented towards industry and agriculture. The corporate income from contract-based commissions had become dominant by the mid-1980s. In the academic research institutes it had become three times higher between 1975 and 1989 (Academy, 1990), mounting up to 55-65%. The gap between research and industry had gradually disappeared. That, however, did not mean real cooperation. We can say that contracted research met real industrial demand in the second part of the 1980s. Unfortunately, these requirements often were too simple; they only meant a service and could not be considered as innovations. Still, they represented the actual level of development of Hungarian industry. That time 45.3% of R&D work was performed by enterprises, 33.1% by institutes and 16.5% by universities (MTA KSzI, 1991) (Table 3).

	Mrd. HUF (curr. prices)		Total % of personnel GDP		Research and engineers (R&E)	R&D personnel per 10000 labor force Total (R&E)
1981	19,4	2,49	51512	22267	102,7	44,4
1982	21,6	2,55	49236	21970	98,4	43,9
1983	20,7	2,31	48740	22132	98,1	44,5
1984	23,0	2,35	49360	22518	99,9	45,6
1985	24,4	2,36	48745	22479	99,2	45,8
1986	27,8	2,55	49148	22974	100,5	47,5
1987	32,5	2,65	47227	22284	96,7	45,6
1988	32,8	2,32	45069	21427	93,0	44,2
1989	33,8	1,98	42276	20431	87,7	42,4
1990	33,7	1,69	36384	17550	-	-

Tab. 3 Total R&D expenditures and personnel (source: OECD, 1993, p.89)

The atmosphere and the rules and regulations did not, and still do not, really favor innovative enterprises in Hungary. We are hardly past the period of the charges of "unfair profit", which threatened the innovative entrepreneurs even at the middle of 1980s. Innovation always requires an invigorating climate and special socio-economic servo-mechanisms. Hostile conditions and the unfavorable climate in Hungary still do not enable the experience of rapidly developing innovative industrialized countries be used satisfactorily.
"... Somewhat ironically" says the OECD (March 1992), "a principal concern of the examining team has been the lack of strategic planning by the Hungarian government to create an environment where objectives are stated and priorities identified. The

concern is heightened by the fact that there is a strong, and perhaps, natural reticence to recognise the new role the government must now adopt." "It will depend on the broad diffusion of professional behaviour" (OECD, 1993, p.40) "In the difficult situation faced by Hungary ... (this lack of professionalism) might well be a recipe for continued decline or stagnation" (OECD, 1993, p.45).

6. THOUGHTS AND CONCLUSIONS

The theory of socio-technical evolution predicts that any innovation is likely to freeze into routine and reproducible activity, which is the precondition and basis of economic performance. Yet the precondition of any routine activity, when the output of total product becomes proportional to the net capital investment (Szántó, 1985), is the first step into the unknown, the change of both the functioning system itself and its environment. The paradigm of economic application and money-making cannot even come into being without the paradigm of ambitious creation and self-development forming the base for it. The greatest benefit of this bifunctional process is not the profit it promises, but the higher level of problem-solving ability the functioning system can reach as a result of its efforts. Functioning signifies goal-oriented problem-setting and problem-solving activity by men and their means, their "success of achievement" (Whitehead, 1966, p.102) of the result they predicted and reached by themselves. The real benefit is the socio-technical evolution of mankind. At the end of the 20th century it must be widely realized, that it is the evolution of the whole mankind that is at stake, when we face the fact of regression on the territory of one sixth of the globe.

The theory of socio-technical evolution also says that innovation, unpredictable and undetermined on the micro- or entrepreneurial level, becomes determined on the macro- or state (region, multinational network, world) level. The role not taken by a government is also a role and, therefore, a determining factor of socio-technical evolution as well. It cannot be replaced by mere free-market self-organisation. Governments of Eastern Europe might implement the policy of ***laissez faire***, trying hard to establish the up-to-date conditions of mere functioning of free entrepreneurship in their countries. This is definitely not enough to stop the regression their societies are in; in its consequences, it will not differ very much from macroeconomic mismanagement. After drastic political changes of 1989, a government policy of ***faire aller***, or strategic orientation of innovative entrepreneurs, is an absolute necessity to bring the process of systemic distortion to its real turning point. It is not just a new amount of money to be pumped in, but the art of economic stability, even when routine production is unbalanced by positive technological changes, a kind of new alliance between the government and the innovative entrepreneur.

With the burden of aggravating economic crisis on their necks, the governments of Eastern Europe seem neither to be able in the early 1990s to get out of the regression by themselves nor to formulate the concept of technology policy and institutionalize its means. The focal point of the current system of values should be gradually shifted from crisis management to innovative entrepreneurship and its ability to innovate. There cannot be simultaneously two focal points of human result-oriented functioning

(Polányi, 1966, pp.55-57). It is also the matter of willpower and intelligence, of course. But the integrating Europe cannot permit itself the risk of negligence and failure. The fierce world-wide competition in innovation forces Europe to integrate. The same process caused the economic failure and disintegration of Eastern Europe. The deepening regression, the process I would call the necrosis of the fabric of innovation of society, or the innovation crisis that part of Europe has found itself in, is the decease of the whole and not that of its parts only; it wil not stop at the administrative borders. Europe as an emerging system cannot isolate itself from its parts. This problem neither can nor should be rendered independent from the faith of mankind.
OECD in its studies is doing its best to show the right course for the governments of East-Central Europe, but it does not seem to be enough. For the time being, in the eyes of Western Europe Eastern Europe is no more than a potential market for its goods and possibly an unwanted competitor. The application of the paradigm of technology policy on the level of an integrated Europe means a must to give up its attitude of naturalistic observation and change toward active goal-setting and problem-solving activity.
Innovation and technology policy is no longer just an inner problem of any European country. The efficiency of parts and the degree of correlation of their interaction effects the performance of the whole. The policy that Europe and its Eastern part desperately need is to avoid the trap of regression and becoming the periphery of the rapidly evolving world.

7. NOTES

1. Organs of financial and legal control are always to serve the stability of the hierarchy and its routine activity. If technological change is not among those characteristics these most conservative institutions of any society regard preferable for hierarchy's stability, then innovative entrepreneurship is not among its values.
2. of employees left laboratories between 1988 and 1990. During the last 3 years the staff decreased by more than 20%.

8. REFERENCES

1. "Az innováció feltételeinek javításával kapcsolatos szabályozási feladatok" (To Improve the Conditions for Innovations by State Regulation), in Hungarian, (OMFB study), 19-8105-Et, June 1982.
2. "Innováció nemzetközi mércével" (Innovation by International Standards), Paprika Z. - Füstös L., Inteam, in Hungarian, (0MFB study), 1982.
3. "Az innovációs folyamat társadalmi, gazdasági tényezôi népgazdaságunkban" (Social and Economic Factors of Innovation Process in Hungarian National Economy), in Hungarian, (OMFB study), 15-8001-T, Jan., 1983.
4. "A kormány innovaciópolitikája", (The Governmental Innovation Policy), in Hungarian, (non-public document), OMFB-IKM-PM, Budapest, 1993.
5. "A müszaki fejlesztés helyzete és feladatai Magyarországon. Beszámoló az Országgyülés Terv- és Költségvetési Bizottsága részére" (The State and Tasks of

Technical Development in Hungary. Report to the Planning and Budget Committee of the Parlament), in Hungarian, OMFB report, May 1988.
6. Polányi, M.: "Értelem" (Meaning), in Hungarian, in "Filozófiai irásai" (Philosophical Writings), 2.v., Atlantisz, Budapest, 1992.
7. Polányi, M.: "Science, Faith and Society", University of Chicago, Phoenix Books, Chicago, 1964.
8. Sándorné, V.: "Piacképességünk tegnap és ma" (Our Market Ability Yesterday and Today), in Hungarian, (OMFB study), 1986.
9. Schumpeter, J.A.: "The Theory of Economic Development", Oxford University Press, Oxford, 1934.
10. Szántó, Borisz: "Fordulópont elôtt" (Before the Turning Point), in Hungarian, (Minôség és Megbízhatóság), pp.5.-11. and 31.-35. Budapest, 1990/5-6.
11. Szántó, B.: "Innováció a gazdaság fejlesztésének eszköze" (Innovation as a Means of Economic Development), in Hungarian, pp.264, Müszaki Könyvkiadó, Budapest, 1985.
12. Szántó, B.: "A teremtô technologia. A társadalmi-technikai evolúció elmélete", (Creative technology. The Theory of Socio-Technical Evolution), in Hungarian, Közgazdasági és jogi könyvkiadó, pp.460, Budapest, 1990.
13. "Science, Technology and Innovation Policies in Hungary", OECD Exeminers Report, Paris, March 1992.
14. "Science, Technology and Innovation Policies, Hungary", OECD, Paris, 1993.
15. "Tájékoztató a hazai tudományos kutatás fôbb adatairól és a teljesitmény értékelésének lehetôségeirôl" (On the Main Figures on Scientific Research in Hungary and on Possibilities to Evaluate Scientific Performance), in Hungarian, (MTA KSzI study), Budapest, 1991.
16. "Tájékoztató a Magyar Tudományos Akadémia gazdálkodásáról és a K+F állami költségvetési támogatásáról" (Report on the Administration of National Academy of Sciences and the R&D Support by State Budget), in Hungarian, MTA Központi Hivatala és KSzI, Budapest, 1990.
17. "Tudománypolitikai irányelvek", (Governing Principles of Science Policy), in Hungarian, Prime Minister's Office, TPB, (non-public document), Budapest, April 1993.
18. Whitehead, A.N.: "Modes of Thought", Free Press, New York, 1966.

THE ROLE OF SCIENCE PARKS IN THE PROCESS OF INNOVATION

MALCOLM PARRY
General Manager The Surrey Research Park, University of Surrey, Guildford, Surrey, UK.

1. INTRODUCTION

There has been a wide public debate on the international stage about the importance of innovation in wealth creation and maintaining a competitive edge in the global market. Those that have participated in this debate in the UK include the initiative Business in the Community, The Government Department of Trade and Industry, (the DTI), such organisations as the Royal Society, The Committee of Vice Chancellors and Principals and my own University, The University of Surrey that in its own right has created an annual public lecture on Innovation.
The debate has centred on the process of innovation, the factors that influence this process and more recently the role of science and research park in the process. This paper looks are current thinking about the process of innovation and the role in which science parks play in trying to effect innovation leading to both the creation of new companies and the growth of existing high technology companies.

1.1 Definition

For the purpose of this talk I have taken as a definition for innovation the one used by the UK Government's DTI:-

> *"the successful commercialisation of new products, services and processes; this distinguishes it from invention which is the generation of a new product idea without necessarily achieving market success."*

In the context of a Science or Research Park the emphasis is clearly on deriving commercially successful products from a science, technology and engineering base. However, there must be a warning which follows from a Royal Society lecture in 1991 (Morita 1992) in which it was stated that "Science alone is not Technology" and "Technology alone is not Innovation". Innovation is a dynamic process and in all economies is essential part of business strategy and development. It relies on good

C. Corsi (ed.), Science and Innovation as Strategic Tools for Industrial and Economic Growth, 85-95.

management of all company functions - i.e. research, design, development, production, purchasing, marketing, sales, distribution, servicing, training, finance and administration.
As a developer of one of the most successful science parks in NW Europe my interest in this debate is the insight it gives me to the market that I supply and the help I give to tenant companies. In broad terms those companies are either small (turn over up to £10 million) or medium sized enterprises (turnover up to £100,000 million) (SMES).

1.1.1 The process
To look at the process it is helpful to briefly review some of the models that have been developed to describe innovation. These models have evolved from the public debate (Mckinsey 1991) and personal experience of dealing with over 100 SMES as well as starting three technology based companies.

1.1.2 Technology driven model
In economic circumstances when product life cycles were relatively extended, compared with the position today, technology driven innovation was not an unusual method of developing new products, processes and services. An example of a successful technical innovation, based on this model, was the "computer mouse" developed in the Research Laboratories of the Xerox corporation. In this instance the innovation or commercialisation of the product was achieved by the Apple Corporation.

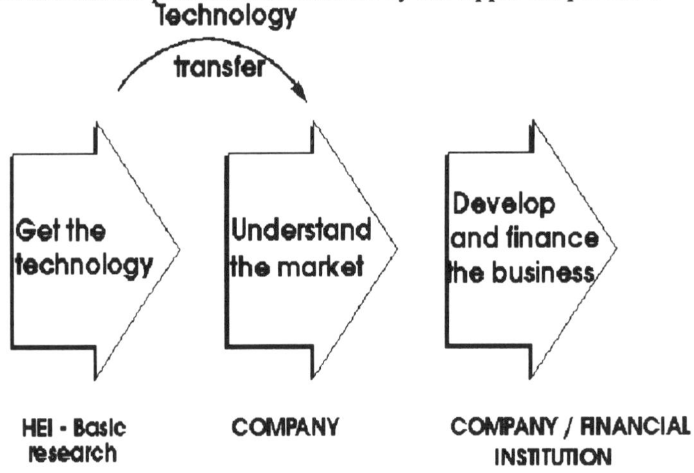

Fig. 1 McKinsey 1991

In this model the research, whether in a University, a government lab or an industrial lab, 'gets the technology' through basic research. The technology, if the business environment is aware, gets transferred into commercial applications that meet market demand. The chain of innovation in this model moves in sequence from laboratory to company and the subsequent involvement of a financial institution before launching the product or service onto the market.
This sequence can be successful; however, it does exhibit several flaws. The most

significant of which are:

- technical advances are rarely whole business concepts, e.g. the development of the liquid crystal display by the RCA Corporation was not a business concept until Japanese watch makers had a need for low power displays for use in their products.
- the process is too slow in a rapidly changing market,
- once out of the laboratory a company may never develop the idea so limiting the process.

Large companies that traditionally separate marketing from technical development often rely on technology push to achieve innovation.

Part of the debate about this process not only cover the relationship between the elements in the chain but also the human sources on how to invoke the right environmental conditions in which innovation can be achieved.

A recognition of the power of the customer has led to some reorganisation of research and development activities. In this paradigm the market place plays a more significant role in the innovation chain.

1.2 The 'Market-Driven' Model

A second linear model relies on commercial companies first defining a market need, it then obtains, from the 'laboratory,' the technology which the company believes will satisfy the market demand. This model gives a greater voice to the customer and is able to react by producing products for niche markets. Again this model has faults which include:

- the risk of losing the market opportunity to a competitor better able to integrate technology and commercial development,
- the market need might well have changed by the time commercial development is complete,
- the technology, although market relevant, might not be affordable and hence not economically viable,
- only stimulating incremental innovation; for example, a customer can highlight an improvement in an existing product but is unlikely to ask for what he does not know is technically feasible.

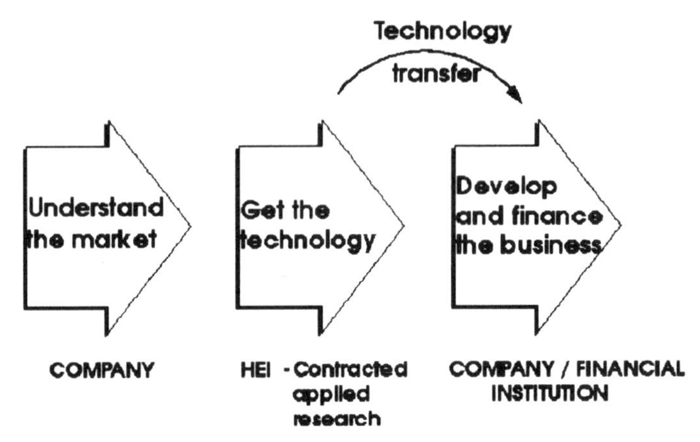

Fig. 2 McKinsey 1991

Examples of innovations which have been market led include the development of the PC. This evolved because of technical capability of the product, high computing power at relatively low cost and size and these factors being combined with the reorganisation of companies IT facilities away from main frame computing. Despite the many successes that have resulted from the linear process of product development further evolution of the innovation chain is now occurring. To accommodate the accelerating product and service life cycle the next step in evolution of the innovation chain has moved towards the creation of an integrated process which involves interaction between the three major parts of the chain.

1.3 The interactive model and its strengths

In both these linear models that describe innovation there are flaws. Those companies that have relied on these two methods of developing businesses that have begun to change their practices to create a new more effective chain of innovation.

One way in which companies are now restructuring, to generate the most effective route to innovate, is to bring together at the outset of the product planning those involved in technology, those that understand the market need and those that understand the commercial side of business development.

Through out the process there is continuous feedback between each of the groups and feedback loops to previous stages in the process of development. In this process their are links into the body of existing knowledge and if this knowledge is not able to resolve the problem then there is an attempt to add to the body of knowledge through research. In some sense relationship between the body of knowledge and the product planning process of innovation can be linear and be either technology driven or market led. It is more likely that innovation will result if the boundaries between the elements that make up the process are dissolved and there is mutual understanding, by those involved, of the elements that make up the process.

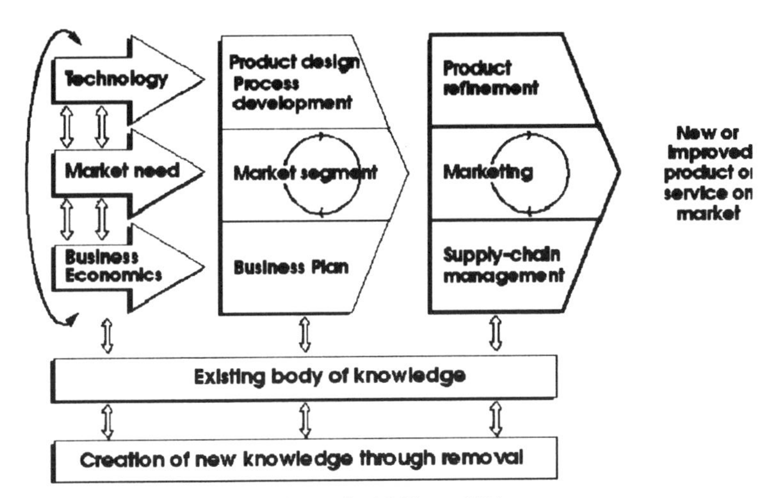

Fig. 3 After McKinsey 1991

Taking this chain apart it is easy to understand that basic scientific research provides information which, though previously unknown, is likely to offer only hints of the future. It is engineers that take these theories and basic building blocks and from them create working technology. Really capturing technology and creating a new product requires a true visionary. These visionaries are people who have a wide understanding of science and engineering as well as a broad vision and true commitment to the needs of society. In essence at the centre of this process are individuals or groups of individuals that are committed to innovation and cross borders between the elements in the process.

Creative or innovative people do not make natural organisational men and women (Adair 1990). Some characteristics which might be displayed by such individuals indicate

- an analytical mind with the capacity to store and recall information
- a high degree of autonomy, self-sufficiency and self direction
- marked independence of judgement
- a novel range of interests
- hard working
- a creative imagination

The ability of such visionaries to be innovative, however, can be restricted by poor management which stifles the process. Despite new ideas invariably coming from an individual it is usually a team that turns this into a commercial success. Understanding this allows for management structures that allow for team creativity which allows the interaction of groups with a different interest to build upon and improve upon other peoples ideas.

Successful innovation is dependent on many factors and judgement is required to create the right mix and match of these; there is no one formula but successful innovation does rely on appropriate management. This must be set up to assist the process by directing the right business environment and ensuring that the company interacts in an appropriate was with the outside world.

1.4 The role of the science park in innovation

There are over 500 science and research parks that have been promoted world wide. Their prolific growth in number reflects an understanding of the need for innovation, the recognition by universities of their role in this process and the interaction of a number of separate trends in the structure of businesses. These trends include:

- an increasing complexity and diversity in the technological base required by any company to maintain a competitive edge; this means companies are turning to external sources of technology and knowledge. These external sources can be, for example industrial laboratories, government research laboratories and universities and of course science parks can facilitate this process.
- the breakup of larger vertically integrated companies into smaller more specialised groups. This has led to increased attention to the creation and support of local small and medium size enterprises (SMEs) especially in high-tech growing industries.

One final pressure driving this process is the need for the science base, in which each nation invests, to show economic returns. Universities have been the beneficiaries of significant research funding from the public sector, which policy makers, not unreasonably, expect to see leading to an enhanced industrial capability.

1.5 Science Parks

Science parks are different from other property based developments in that they :

- have formal operational links with a university or other higher educational institution (HEI), or major centre of research.
- are designed to encourage the formation and growth of knowledge based businesses,
- have a management function which is actively engaged in the transfer for technology and business skills to the organisations on site.

It must be emphasised that ultimately Science Parks are about business, they are about commercialising science by whatever route possible, they attract a wide range of different businesses involved in a wide range of technologies, and these businesses are run by people with varying levels of skill, comprehension of the market and vision.

1.6 The role of science parks in the process of innovation

They review of the models of innovation gives some insight into how science and research parks can help facilitate innovation; in the rest of this talk I want to be more specific about this relationship.

Science parks create the right physical environment in which innovation can be generated. To an extent this physical environment reflects cultural norms and tastes; however, the buildings must be made to be flexible in use and be available on the right kind of occupancy contract to accommodate those involved in creativity. The location must be right in relation to the markets addressed by tenant companies and

communications to these market must be effective. Without an involvement in the market innovation is more likely to fail.
Science parks must give the right image for companies that locate on them. The words science park has already acquired a meaning in international business circles. In general this meaning reflects a good image and comes partly from the fact that science parks tend to attract entrepreneurial companies that have a natural inclination to innovation.
Science Parks must develop the right intellectual environment to help entrepreneurs gain the most from the resources of the host institution including business skills education, access to equipment, staffing and ideas. Experience has identified the main processes by which scientific and technological knowledge is exchanged with knowledge form industry. These include:

- Research links and collaboration, whereby a host is asked to generate new forms of information, or carry out specific research tasks,
- Information transfer in which existing knowledge or information is passed on in a formal sense through some form of learning experience
- Knowledge transfer via personnel mobility in which there is movement of staff or students carrying with them formal or tacit knowledge,
- Spin-outs and externalisation, where a body of knowledge in an host organisation is transferred to the private sector or some other organisation by the wholesale transfer of a group of people over an organisational boundary.

The pattern of relationships between academics and tenant companies will naturally vary with the size and technological sophistication of tenant companies. Larger tenant companies are more likely to have the resources to commission research and become more deeply involved with a park's host. Smaller companies tend to rely on consultancy services as part of a development, which is a transfer of information from the existing body of knowledge rather than the creation of new knowledge through research. However, whatever the size of the tenant companies, it is important that the pool of visionaries can take forward innovation increases the borders between science technology, engineering and market knowledge and this border is progressively broken down so that a innovation orientated culture develops.
The process of innovation is fundamental to the growth of high technology companies. As a developer of a University related Science Park, the Surrey Research Park which indicates in its mission statement a commitment to assist in economic development and the growth of high technology companies, the process of innovation of formation of companies and company growth are of interest.
Effective innovation helps lead to economic development. in turn economic development relies on the growth of economic activity by either existing or new companies.

2. A MODEL OF GROWTH OF A HIGH TECHNOLOGY INDUSTRIAL SECTOR.

There is very little quantitative data available about the creation and growth of high technology companies so any analysis must rely on experience. Based on my own experience over the last 10 years I have concluded that the most realistic model describing how high technology companies are generated and grow is that of Miller and Cote (1987). An interpretation of the model in the light of experience of developing The Surrey Research Park is set out below.

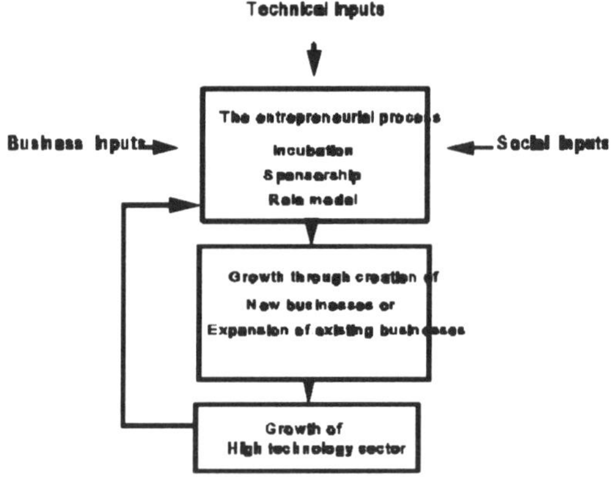

Fig. 4 Miller and Cote 1987

3. THE PROCESS OF GROWTH

Growth occurs when there is an addition of new economic activities to the existing economy. This may result from either starting new businesses (start up/spin-off growth) or expanding existing businesses (corporate growth).
Some of this growth occurs when more of an existing supply is created to match increased demand. Although the growth of some companies on science parks falls within this category the main intention of any science park is to enable growth through innovation in technology and its subsequent application to new products, processes, services and methodologies.

4. SCIENCE PARKS AND START UP COMPANIES

Science parks play the important role of providing, for business start up, appropriate space, business services and access to the full resources of the host institution. My experience of "start up" companies is that they are usually associated with younger industries that are in a cycle of innovation and growth during and demand for their products and technology exceeds supply and this gives the entrepreneur the opportunity to build a new business.

5. SCIENCE PARKS AND CORPORATE GROWTH

The markets in whcih larger companies operate are changing significantly. It is clear that these are becoming more specialised, their customers have more sophisticated requirements and quality and choice are becoming more valued that quantity and uniformity. To meet these demands large corporation have begun to increase their proactive of breaking up into smaller business units and forming alliances with small firms. universities and other centres of excellence to acquire the technology to match the demands of the market, i.e. turning to external sources of technology and knowledge (Miller & Cote 1987)

The development of science parks with their flexible policies for letting and management of space and concern about business support and technology transfer, is an attempt to satisfy the needs of these changing businesses and to help them develop technology strategies.

These technology strategies include:

- developing links with small firms with specific technical skills;
- using research grants to develop ideas;
- employing contract research organisations;
- creating joint ventures to develop target technologies;

6. HE RIGHT ENVIRONMENT FOR GROWTH

At the heart of growth is the entrepreneur, ie. "the business organiser or manager that sits between capital and labour". However, if these individuals are to reach their full potential in the high technology sector it is helpful if the business environment has certain characteristics; these include:

- the existence of "incubator" companies,
- a supply of companies that offer s sponsorship and
- the existence of successful role models.

manufacturers to pharmaceutical companies and include service companies such as stockbrokers, banks and insurance companies. Their dependence on technology means that they act as sponsors and stimulate the creation of new businesses by offering contracts to innovative technical companies. Finding these commercial sponsors in the local economic fabric is critical to the success of start ups.

In Surrey local sponsors include: the Civil Aviation Authority, City of London based financial institution, military contractors and government defence procurement programmes.

7. SUCCESSFUL ROLE MODELS

Exposing would-be entrepreneurs to the success of other start up businesses helps to accelerate the rate of business creation. The is no doubt that success breeds success. We run meeting where successful entrepreneurs give talks about their own companies and experience. Our park has become recognised as a friendly place to start to talk about a business if someone is thinking of taking that first vital step.

Incubation, sponsorship and role models exist as opportunities in the business environment; however, to pursue these opportunities it helps if, in the business environment, the relevant infrastructure also exists.

8. BUSINESS INFRASTRUCTURE

8.1 Business inputs:

The availability of good managerial advice and access to funding can accelerate the development of high technology companies. Clearly how this is arranged and accessed is a matter of management policy. Some parks offer venture capital funds to assist in this role;

8.2 Technical inputs

For the growth of high-technology companies there must be a driving force of several local sources of leading edge technology. These sources not only act as sponsors but also incubator companies. Examples include:

- advanced laboratories of technology-based firms that conduct market driven research;
- contract research organizations that undertake projects for private firms and government agencies;
- universities;
- and finally government laboratories.

8.3 Social inputs

The prevailing social and commercial conditions must reward entrepreneurial activities. Examples of the right conditions include:

- rewarding creativity, invention, innovation and entrepreneurial spirit;
- offering a market for start-up firms by employing them as government contractors or as suppliers to the defense or space industries before shifting their client base towards the commercial market.
- setting tax regimes that help growth.

Science Parks must develop the right business environment by facilitating a relationship with financial institutions, those with business skills and young people qualified in the right technologies. The right manager will create a feeling of trust with both financial institutions, the local community and the host institutions staff and students.

Science Parks must have the right management structure to ensure that each company can get what it wants and needs from the park and its host institution including access to the right kind of training programmes to help businesses development.

Science Parks compared with other property initiatives provide this opportunity because of the flexibility they offer to entrepreneurial groups that need to work together for short periods of time, access to knowledge and if the park has the right image it will attack those companies that are innovative by nature.

However, regardless of a science park the whole process relies on creativity in technology; creativity in product planning and creativity in marketing which are fostered by creativity in management.

9. REFERENCES

- Morita,A. 1991, S does not equal T, and T does not equal I. The First United Kingdom Innovation Lecture, Royal Society, London. DTI/Put 90K/30K/2/92.
- Mckinsey, 1991, Partners in Innovation, report for the Prince of Wales Award for Innovation, London.
- Miller,R.and Cote, M. 1987, "Growing the next Silicon Valley", A Guide for Successful Regional Planning, Free Press ISBN 0669145777
- Adair,J. 1990, The Challenge of Innovation, Talbot Adair Press ISBN 09511835 32.

ASSOCIATION FOR THE PROMOTION OF SMALL INNOVATION ENTERPRISES, TECHNOLOGY CENTRES AND TECHNOPOLISES, MOSCOW - RUSSIA

EUGENIO CORTI
President of CALPARK SpA, Science & Tecnology Park of Calabria Italy, Vice-President of Technologies for International Technoparks (TIT) Ltd. St. Petersburgh, Russia

1. TECHNOLOGY AND TECHNOLOGICAL INNOVATION IN REGIONS WITH A CONSISTENT DELAY OF DEVELOPMENT

Several scientific institutions are located in Russia (i.e. Universities, Research Institutes and Centres, Military Industrial Research Centres, Large Innovative Companies, new High-Tech companies, etc.), they manage a very large number of very qualified scientific and technological laboratories with thousands of scientists and technicians, where they produce a great amount of very interesting scientific results. Most of them are at a very high level according to the international standards.

On the other side it is quite easy to verify by a foreign visitor that the places where the above mentioned scientific results might be applied (i.e. all the industrial sectors, agricolture organizations and companies, the services sectors, the Public Administrations, etc.) constitute a very weak system, with several companies and organizations which are not able to compete on the international market, and are not able even to respond to the local market.

Therefore it is almost obvious to say that the connecting links between the two above worlds are dramatically weak. In Russia most of the interesting scientific results are not transformed into useful technologies, because the demand from the industrial and services system is generally speaking very poor.

In the market economy the Region where the linkages between the places where the technical and organizational knowledges are produced and the places where those knowledges might be applied are weak, i.e. the Region where the diffusion and the transfer processes are very uncommon, is called "Region with a consistent delay of development".

To analyse those kind of Regions it is necesary to consider any single organization inside the specific economic sector, which is interested to acquire a new technical and organizational knowledge, in order to understand the real conditions to operate any diffusion-transfer process.

C. Corsi (ed.), Science and Innovation as Strategic Tools for Industrial and Economic Growth, 97-123.

Let me concentrate to the industrial sectors, where a single company operating is trying to survive by increasing its competitive capacity.
It is well known that the general objective of any single company in any industrial sector, acting inside the market economy, must be *"To Satisfy the Clients and to Increase its Own Profit "*, and the powerful tool for this company to perform its general objectve is the patrimony of all the Technologies the company possesses.
But the management of that company might ask the following questions:

- Which Technologies must be selected inside its own Technology Patrimony, in order to reach the General Objective ?
- Which Technologies might be acquired from outside in order to encrease its own competitiveness?
- How to manage them ?
- Which human resources and financial resources must be used ?
- Which results may be expected ?
- Which costs may be accepted ?

To answer to those questions it is necesary to recall the definition of technology, which I propose to be the following: *Technology is the set of all necessary omogeneous technical and organizational knowledges, by whose use, the company, which possesses them, may realize its general and operative objectives .*
With the above definition I mean that *technology* is synonymous of *information* , and therefore *technology diffusion* is synonymous of *information flow* .
Inside a company other two resources are present besides *technology* , they are *the financial resources and the market resources* . To manage individually the three resources the company needs three organizations, but also it needs organizations to manage the interactions among the above three resources. However to manage organizations means to manage human resources, therefore the innovative company puts a great effort to manage people.
A single company has to offer to its clients a *"Bigger Value "* to perform its general objectve, which means that the company has:

- to select which products must be offered to which clients;
- to consider carefully the changes in behaviour and needs of the clients (DEMAND) and the changes of the contents and the possible uses of the familiar technologies and those of new technologies (OFFER);
- to cosider new strategic and organizational trends, new approaches and new management tools.

One single company is able to offer a *"Bigger Value"* only if it is able to manage innovative processes inside its organization. In Fig. 1 a picture of the innovative process is reported, where the actual situation of the company and its environment must be well known to the company management as well as the alternatives of possible future situations. To transform the actual situation into one of the future possible situations the company must organize suitable strategies, which implies the use of suitable resources, among which the human resouces will perform suitable activities, and it will use, develop and change technologies.
Le me be more careful with the meaning of the *technological change* . I'm proposing

the following definition: *"Technological Change is the modification of the olevel of the use of a particular technical or organizational knowledge (i.e. of a technology)".* Therefore the definition of the technological innovation is the following: *"Technological Innovation is a technological change which is performed to reach a predeterminated objective ".*
The above two definitions are very important because it appears that their meanings are very different: the second definition is a small subset of the first one, i.e. among many possible technological changes, there are only few that become technological innovations, because only few of them will curry out appraisable results.

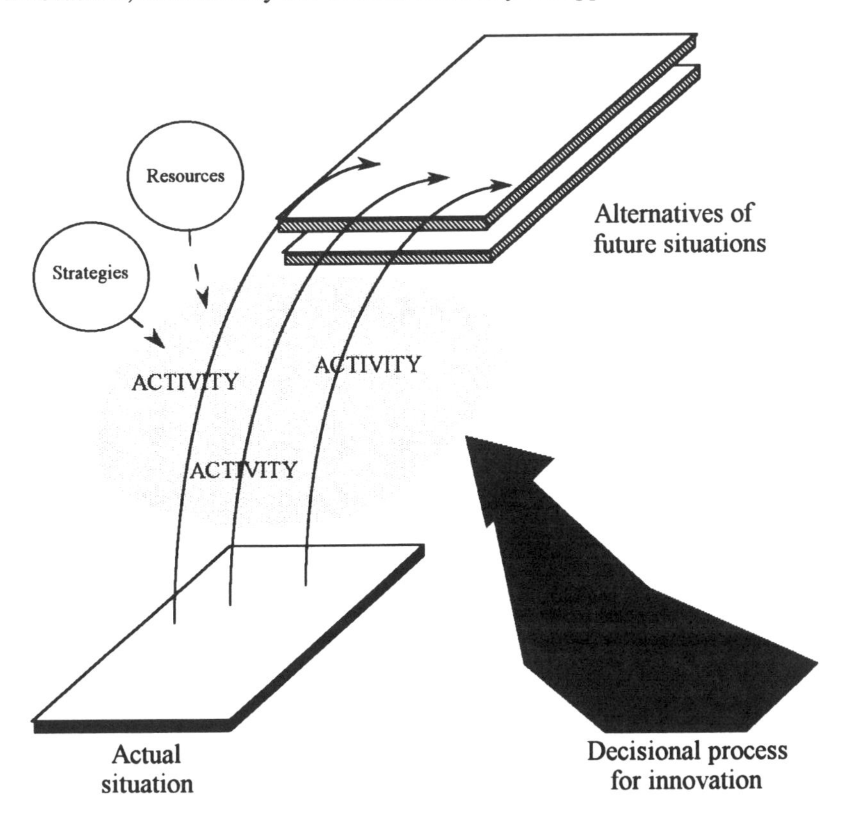

Fig. 1 The decisional process for innovation

The relations between the general objectves and the technological patrimony must be tight, and the company may have two directions:

1.1 to make more attractive either the actual products or the new ones:

- by improving the performance characteristics, as seen by the market, of a selected set of products (actual and new ones);

- by decreasing the direct costs of a selected set of products, in order to decrease their prices;

thus the company has to identify the opportunities to innovate the market relationships. Among these relationships the company has to identify the priorities of the so called "Technological Innovations of Products ".

1.2 to make more efficient the organization of the company:

- by reducing the direct costs of a selected set of products, and keeping constant the performance characteristics;
- by reducing the indirect costs of the selected organizational structures of the company;
- by reducing the costs due to all kind of the non-quality costs;

thus the company has to identify the opportunities to innovate the efficiency of the company, among which the company has to identify the priorities of the so called "Technological Innovations of Processes ".

In Fig. 2 a model of the management of the technological innovations is reported, and we can see that the model is made by two levels: one is the analysis level where the company has to make the analysis of the demand and the analysis of the evolutions of the technologies from the environment, and at the same time the company has to make the analysis of the level of the efficiency of its organization and the analysis of the internal possibility of the technological change.

The outputs of the analysis level will be the opportunities to improve the relationships with the market (i.e. market opportunities), the possible technological changes inside the company and the opportunities to increase the efficiency of the whole company or part of it (i.e. efficiency opportunities).

The above opportunities enter into the Synthesis Level of the model, where now the company has to compare the market opportunities with the possible technological changes to find the technological changes that allow the company to reach that market opportunities (product technological innovations). At the same time the company has to compare the efficiency opportunities with the possible technological changes to find these chnges that allow the company to reach that efficiency opportunities (process technological innovations). Those technological innovations will be evaluated and selected and therefore the company will have the priority intervenctions for technological innovations.

A possible model of the relationships between the company and its environment is reported in Fig. 3. In this figure the following resources for the company are drawn:

- financial resources, which may include the sales of the products of the company as well as the financements from the partners of the company and from the banks;
- material resources, which is needed by the company in order to manufacture its products;

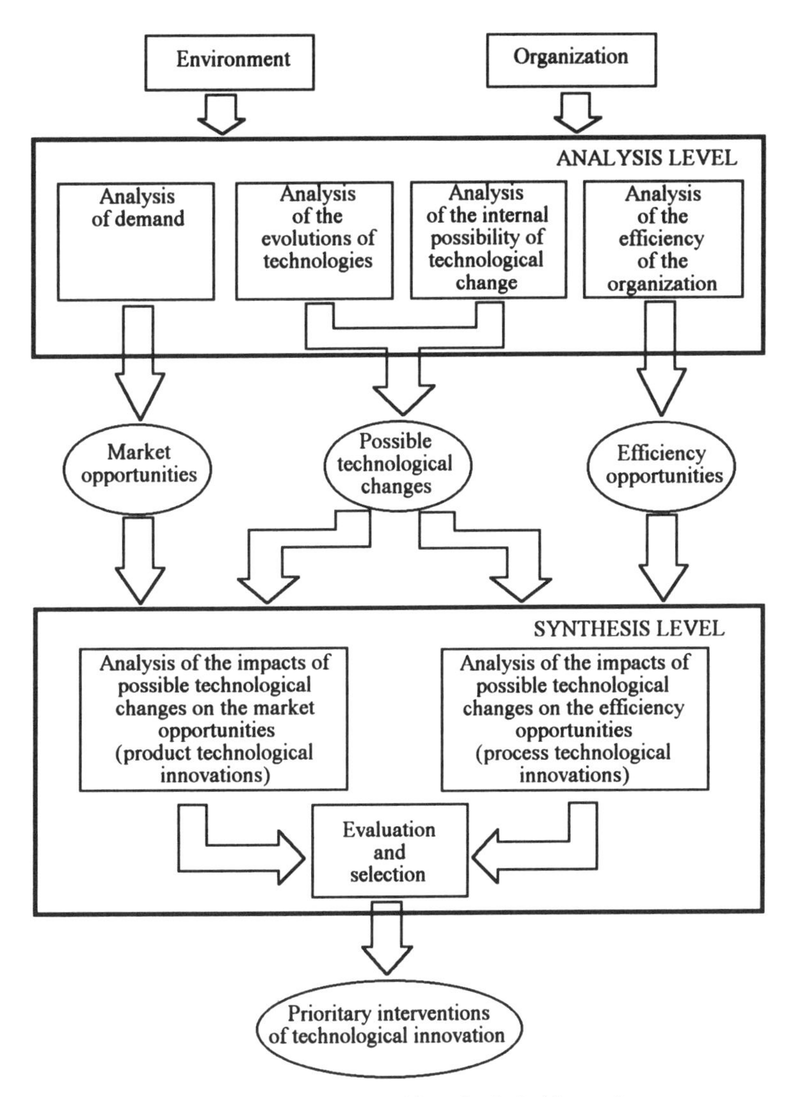

Fig. 2 Model of management of the technological innovations

- human resources, which include not only the new eventual employees but also the actual employees after a suitable training;
- information resources, which include also, as I have said before, all the needed technologies to manufacture the products;
- environmental resources, which are related to the environmental conditions, such as the quality of the transportation resources, of the residential resources, of the other social services (schools, hospitals, etc.), which make the life outside the company more valuable. It is important to recall that the quality of life outside the working place conditions the quality of the innovative processes.

The above resources in the model reported in Fig. 3 enter into the company, where, on the basis of the already approved strategic plan, are used at the right time and transformed in order to obtain the outputs of the company.
Notice that every resource will stand sometime inside a conceptual warehouse, and time to time it is taken away to be used to manufacture products. This implies that if the time of standing inside the warehouse is too long, the value of the resorce will be dissipated, as shown in Fig. 3.
The outputs of the company are sold to the users, and the success of this sale is closely related to the level of the utility and convenience perceived by the users, in such a way that in case the level is high the users want to acquire those outputs, on the other hand in case that level is low the users are not interested to acquire those outputs. The perceived level of utility and convenience is influenced by some external variables, which depend on the general conditions of the local environment.
The model tells us that the level of the perceived utility and convenience influences the flow of the resources entering into the company. In other words the valves, drawn in Fig. 3, are open only if the level of the perceived utility and convenience appears to be high, but they are almost closed in case such level is quite low.
It shows very clearly that any company in order to get success has to offer to its clients products (outputs) which cause in them a high level of the perceived utility and convenience. On the other hand, in case the company understands that the level is too low, it must do somethink to change the situation. The only way to change the situation is to change the outputs of Fig. 3. In this figure every output is characterized by three attributes:

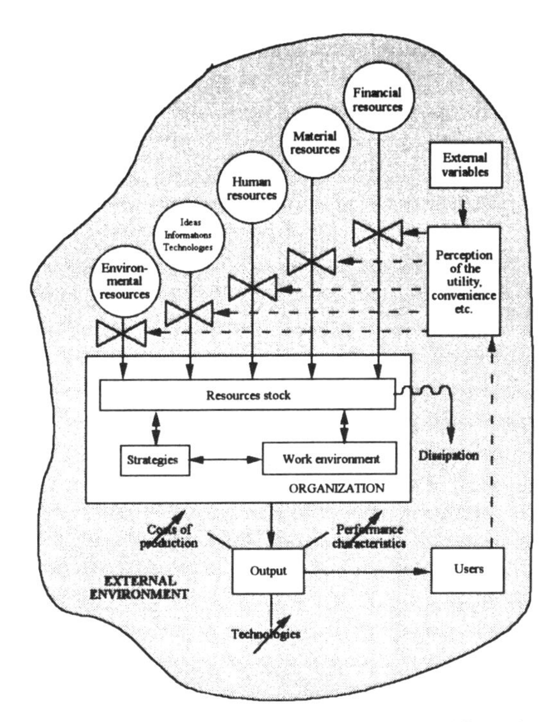

Fig. 3 A model of the relationships between the company and its environment

- the performance charateristics as seen by the users and not by the company,
- the sum of all the elementary costs necessary to manufacture that output,
- the whole set of technologies (technical and organizational knowledges) needed to manufacture that output.

The first two attributes may indicate the objectives to change (why and what to change), the last attribute indicates the tool the company has got in order to be able to change (how to change). In other words the company may be forced by the market to decide to change some of its products in such a way that the changed products will be more acceptable by the market. At the same time the company may decide to change some activities inside its organization in order to increase the efficiency, which means to reduce some costs in the company and therefore it is possible to reduce the costs of a single product.

The model reported in Fig. 3 explains very clearly the role of both the product technological innovation and process technological innovation, in fact to change the suitable performance charateristics by means the suitable technology changes is a typical product technological innovation, while to change (to decrease) a suitable set of costs by encreasing the efficiency of the part of the company is a typical process technological innovation.

To perform the above mentioned technological innovation processes the company has two operative directions:

A) a better use of the technologies already used inside the company,

B) acquisition of new technologie from outside the company.
As far as it may concern the first direction, the company has to improve its capability to manage technologies, i.e. to manage knowledges and information, therefore to manage pleople. A company which has an high capability to manage technologies is an innovative company.
Notice that an innovative company is a company which remains competitive in time and space.

2. THE DIFFUSION AND TRANSFER OF TECHNOLOGIES

Here in this workshop, due to the lack of time, I would like to describe only the second direction, which may be represented by the Fig. 4, where it is explained that the scientific results, made by the organizations which produce new technical and scientific knowledges, such as: Universities, Research Public Centres, Industrial Research Centres, Industrial Development Centres, etc., are first transformed into new applicable technologies.
The transformation from scientific results to applicable technologies is a quite complex process, which is usually made by the most innovative organizations which are able to evaluate and select the scientific results to find out the technical possibilities to trasform them into technologies useful for their aims. Recently this role has been enphasized by the Science & Technology Parks, expecially in the Region with a consistent delay of development.
After that trasformation the new technologies and the actual technologies are ready to be diffused among the organizations, which may use them, and transfered, as shown in Fig. 4, from the technology source into the single organization by means a *"technology diffusion & transfer process"*. One important result is the following: such process, which is a quite complex process, easily accurs in a short time when the organizations which are interested to acquire such technologies are enough innovative. On the other hand if the organizations are less innovative, generally speaking they are not able to acquire technologies at a reasonable time delay without any help from outside.

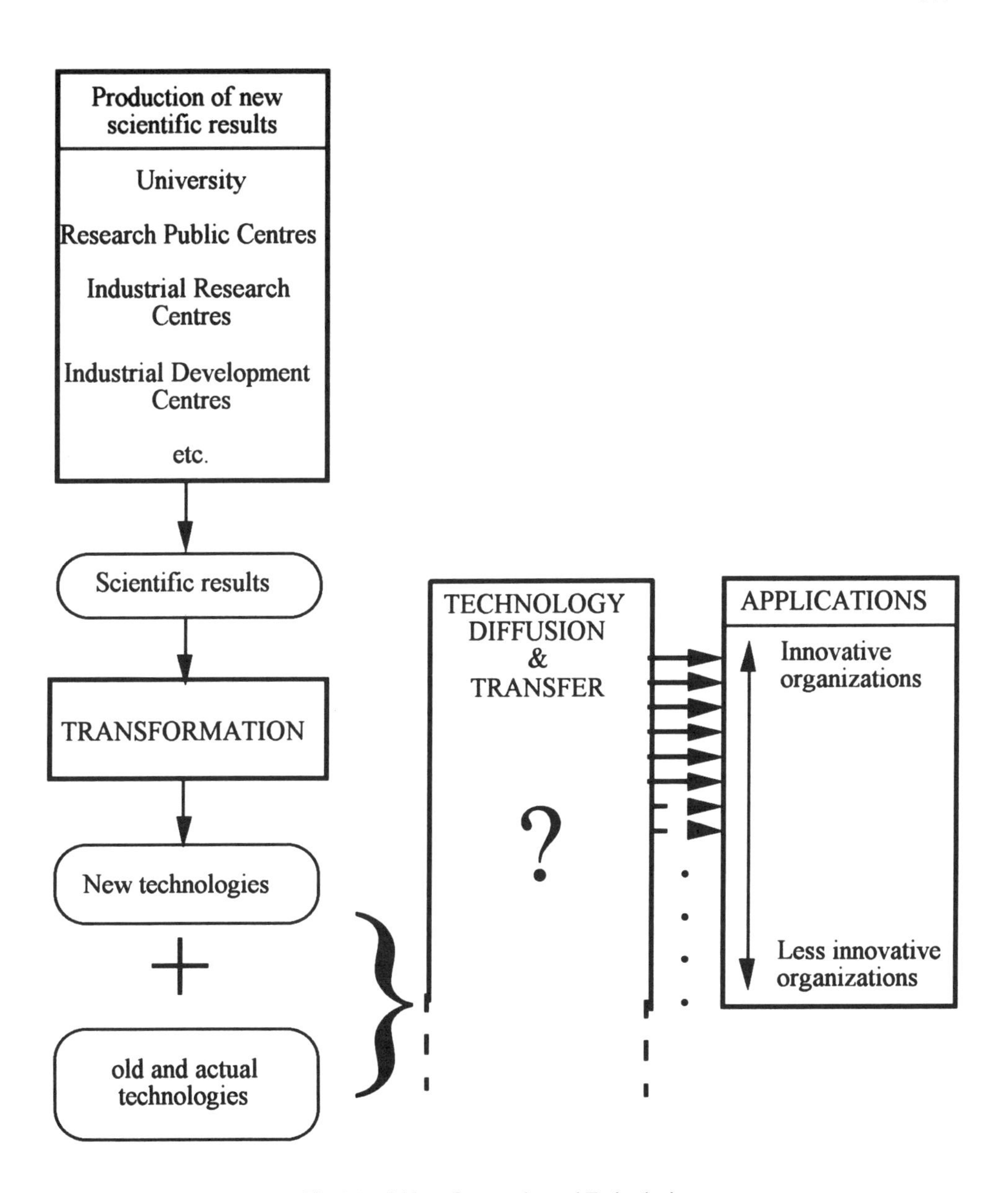

Fig. 4 Aquisition of new and actual Technologies

Again in a Region with a consistent delay of development, since the density of innovative organizations is very low, this process must be performed by suitable organizations, such as the Science & Technology Parks.
The *"technology diffusion & transfer process"* takes a different time interval according to the level of innovative capability of the company, besides, in case for any reason one company wants to decrease this time interval, some barriers to innovation will appear:

shorter the time interval, higher the barriers to innovation.
The main problem for a Region with a consistent delay of development is not to-day to produce new knowledges but to diffuse, to transfer and to use in more efficient way the already well known technologies. Thus we need to understand better what the diffusion process, the transfer process and the use of such technologies mean, in order to understand how to design effective external organizations, such as the Science & Technology Parks, which may play the same role of the innovative organizations in the higher developed Regions.
First of all let us consider the rate of growth of the diffused information in the world vs. the years, and we will get a curve which is growing alwayes faster in the last years, as it is reported in Fig. 5. This dramatic curve shows that the quantity of the produced technical information every year is getting larger and larger.
The diffusion process inside a social system may be represented by a linear model, which implies that:

1) if one organization of the social system gets the right information, it accepts the innovation;
2) any organization is indipendent from any other organization in the social system.

However the linear model has been contradicted by many experiences. The most well known case is the one of the diffusion of the hybrid seed of the mais in the 1930' years into few american regions (1928-41).
The diffusion of this new seed, started with a complete promotion among the farmers, based on the fact that the productivity per hectar should have been increased by 30% by using the new seed.
The results were impressive:

- 14 years were needed to complete the diffusion, while the time by which half of the farmers adopted the innovations has been 9 years;
- the behavior of one single farmer has been different from the behavior of the other farmers;
- the informal communication played an important role;
- the critical role of "Opinion Leader" came out;
- the process of innovation diffusion may be accelerated by introducing suitable diffusion structures, such as:
  - innovation agents;
  - experimental activities;
  - demontration activities.

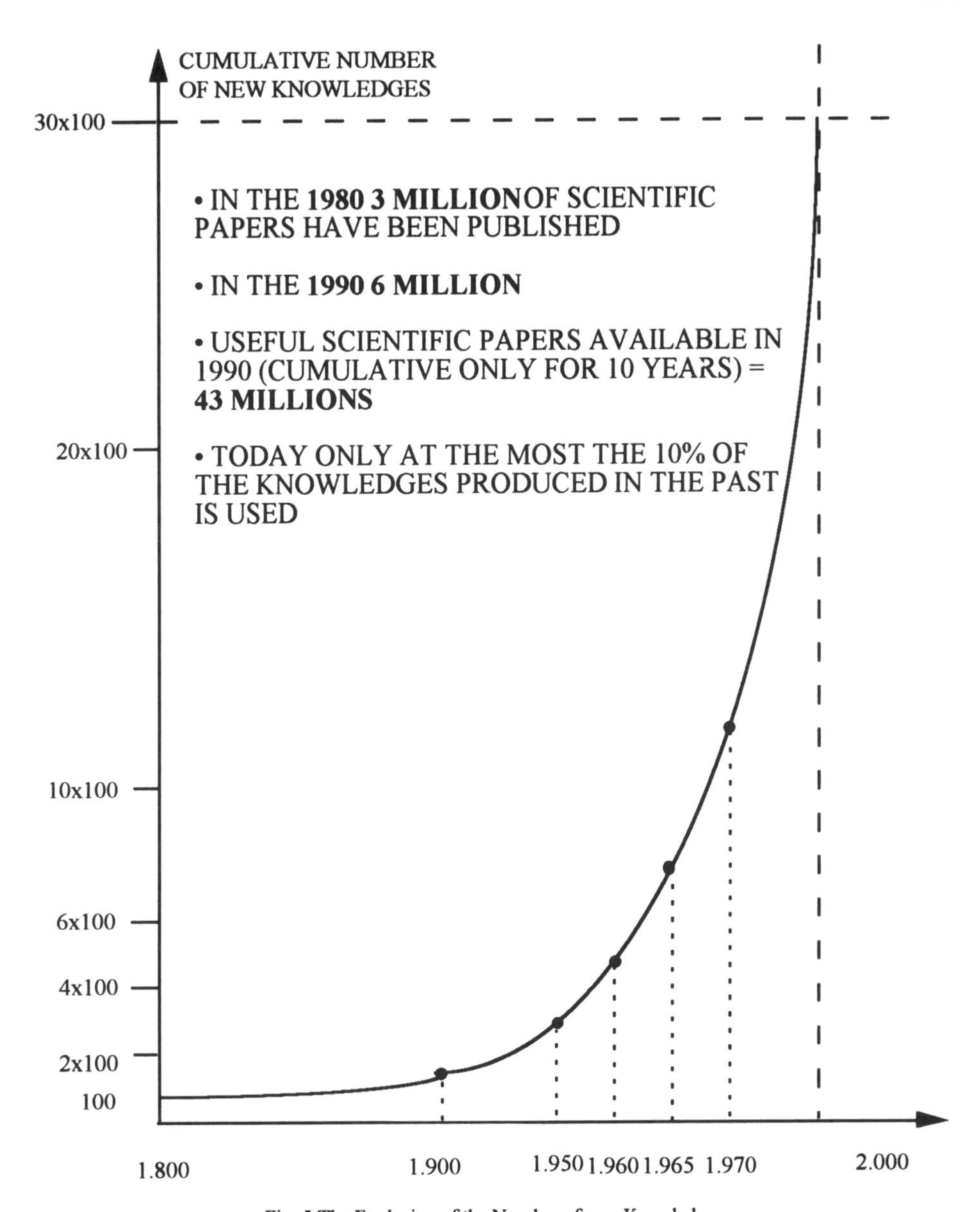

Fig. 5 The Explosion of the Number of new Knowledges

These results made clear that the linear model cannot be representative of the real situations, and then the non-linear model has been proposed.
The non-linear model, represented in Fig. 6, where S is the source of innovation and Ai is the ima single element of the social system, describes much better the real situation.

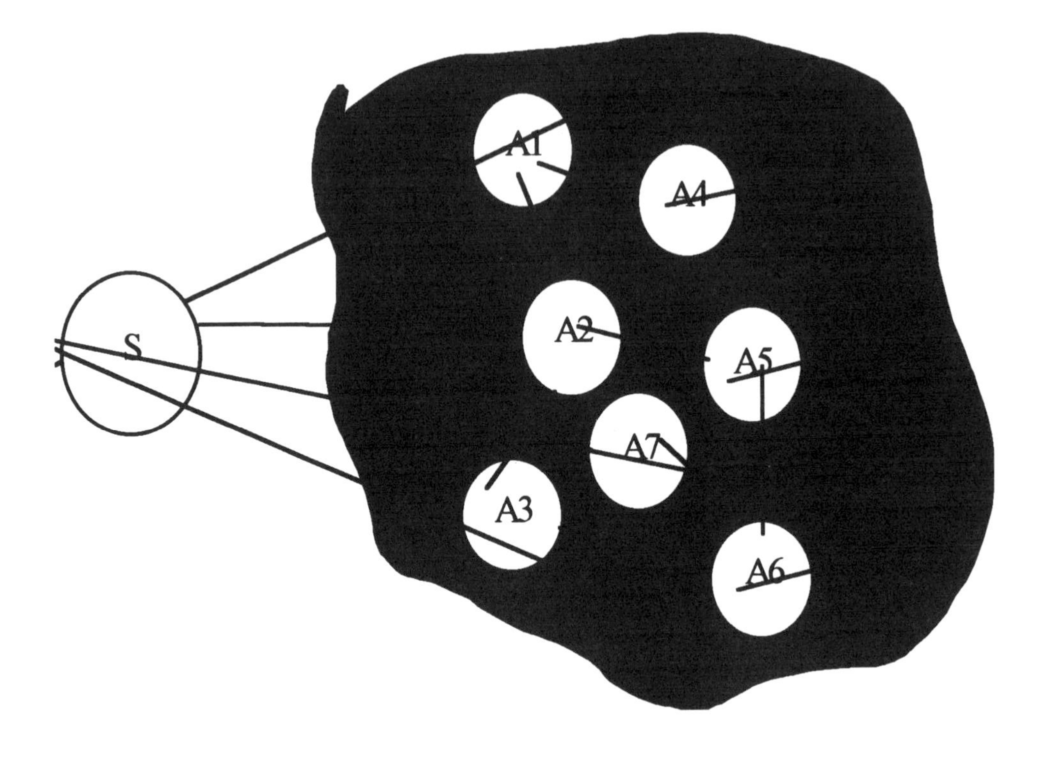

Fig. 6 The non-linear model of the diffusion process

The characteristics of the non-linear model are the following:

1) the process of innovation diffusion is not a direct process, but it implies the indirect flow of information, i.e. several members are informed by other members rather than by the information source;
2) the information flow encreases if the informal communication among the organizations is high;
3) therefore the probability to get the right information at the right time encreases if the density of the organizations is high.

One possible interpretation of the non-linear diffusion process is the adoption curve, represented in Fig. 7. In any social system the time reaction of the single member of the system has a different lenght, therefore the members of the system distribute themselves along the time axis, as shown in figure.

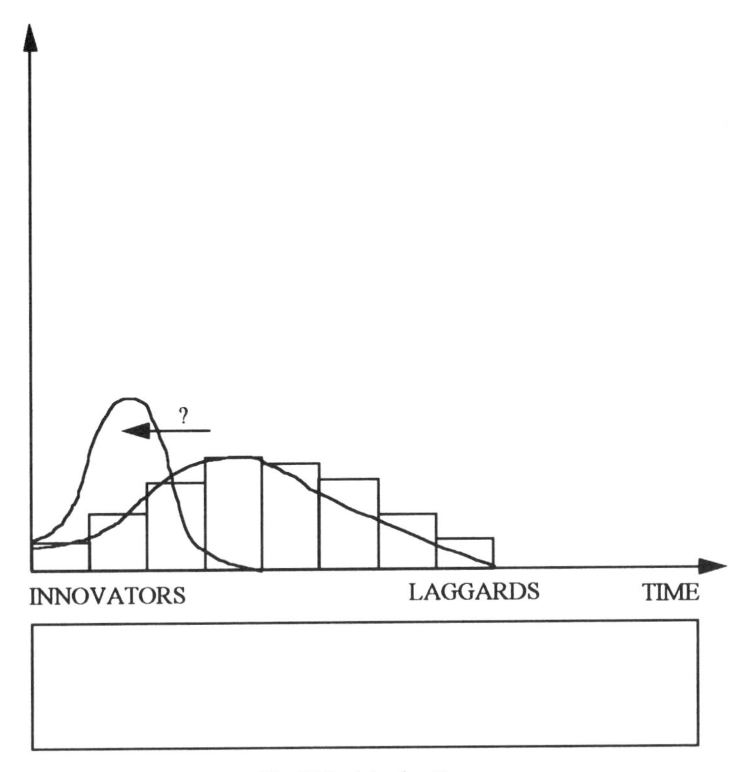

Fig. 7 The Adoption Curve

When the time needed by members of the social system to decide the adoption (or refusal) of the innovation is short, those members take the name of innovators . Some members make the decision to adpt (or to refuse) very late, those are called laggards . Most of the members are settled in the middle. It is interesting that this behavior pf the adoption curve is very general: it applies also when the social system is a large R&D laboratory.
The main problem for every social system involved into a diffusion process of a innovation is to push the adoption curve to the left. But to understand better how to do it, it is useful to examine closely the adoption mechanism. The first step is the innovation decision process, by which the single member makes the decision to adopt (or to refuse) the given innovation.
The model of the innovation decision process is reported in Fig. 8. Where the total adoption (or refusal) time is equal to $T_1 + T_2 + T_3 + T_4$. The first phase of the model is *"knowledge"*, and this phase is influenced by the formal communication. Any individual spend a different amount of time for this phase, but this phase is not enough to make a decision, therefore after that phase the model shows the phase "persuasion ", which implies that the individual asks himself about the utility to make an adoption or a refusal to be convinced that innovation is really interesting for him. Notice that this phase is influenced by the informal communication and not by the previous formal

communication.

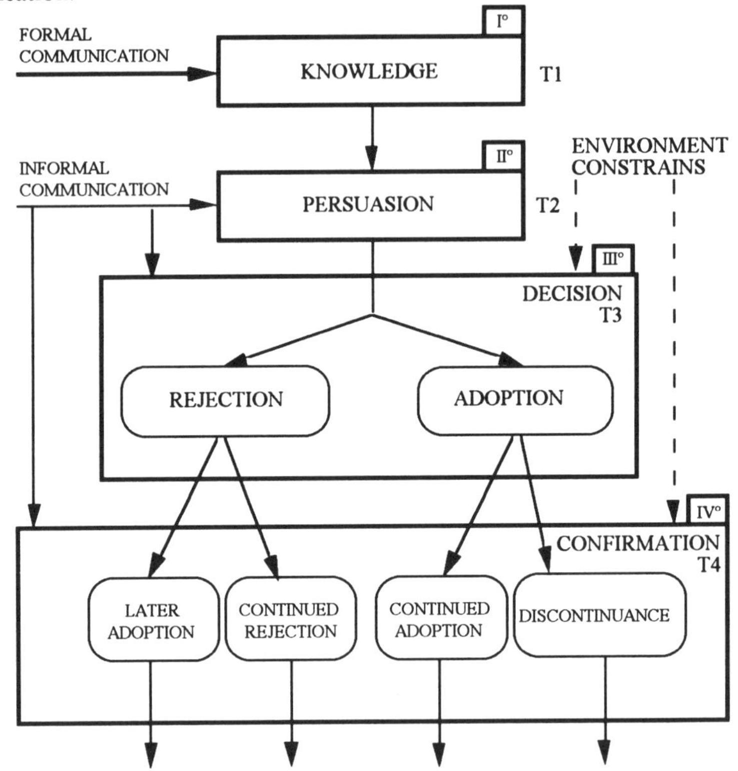

Fig. 8 The Innovation Decision Model

After the phase *"persuasion"* the individual believes to be ready to decide, i.e. he enters in the thrird phase *"decision"*, where he may either adopt or refuse the proposed innovation. At the last it is necessary to consider the phase *"confirmation"*, which is very important because some times the individual makes the decision and then he wants to change his mind.

Notice that only the first phase is influenced by the formal communication, but the other three are influenced by the informal communication.

The Technological Innovation Adoption Process may be obtained by the sum of the following more simple processes:

= Identification and Selection of a New Technology (Innovation) +
+ Innovation decision process +
+ Technical and Organizational Redesign of the their own structure +
+ Technology Transfer Process; +

where the *Technology Transfer Process* means a real acquisition of a new technology, which takes care of the following operative aspects:

◊ Technical aspects;
◊ Organizational Aspects,

◊ Legal aspects,
◊ Economical aspects,
◊ etc.

In the case of a Region with a consistent delay of development, the Technology Transfer Process interests a very small number of organizations. In Fig. 9 a matrix of four cells is reported, where the rows are the services offered to the local organizations, divided into Technology Transfer Services and Organizational and Managerial Consultancy Services. The columns are related to the explicit and latent demand of services from the single organizations in the Region.

Every organization in the Region will occupy one cell, as follows:

◊ CELL A: very few organizations know their problems and explicitly demand new technologies to solve their problems;
◊ CELL B: few organizations demand general consulting support to solve their problems, which can be solved without using new technologies;
◊ CELL C: most of the organizations in the Region are located in this cell, perhaps they know their problems, but they don't know how to solve them and what to ask to get support;
◊ CELL D: few organizations are located in this cell, they really have some problems that may be solved with a new technology, but they don't know which technology and even they don't know how to ask for.

It is interesting to notice the kind of services that the organizations in the cell A and cell D are looking for. For cell A they may be summarized as follows:

to support the definition of the problem inside the company and the possible technological solution;

	SERVICES DEMAND	
	EXPLICIT	LATENT
TECHNOLOGY TRANSFER SERVICES	A	D
ORGANIZATIONAL AND MANAGERIAL CONSULTANCY	B	C

Fig. 9 Technology Transfer Services in a Region with a Consistent Delay of Development

◊ to evaluate and select the suitable technology;

◊ to organize the experiments and the demonstrations with the selected technology;
◊ to perform the Feasibility Study (including: where to find that technology (technology source), which are the acquisition conditions, which inpacts on its own organizational structure, which advantages with that new technology, where and how to get the appropriate funds, etc.);
◊ to support the negotiations with the technology source;
◊ to formalize a purchase contract (including: transfer conditions, payment conditions, tests and inpections conditions, etc.);
◊ to support the implementation of the acquired technology (including: the redesign of the organizational structure);
◊ to support the tests of the new technology.

For cell **D** they are:

- before offering new technologies, even if those new technologies may solve the identified problems, it is necessary to organize a sensibilization program for the companies located in this cell, by using demonstrative and experimental actions, and then, when the latent demand becomes explicit, the organization will be active in cell **A**..

The Technology Transfer Process may be represented by the flow diagram reported in Fig. 10, where we can start to describe the process from the member of the social system which has not yet adopted the innovation because of the natural resistance to the innovation influenced by environment constrains. But in the phase of evaluation of the innovation this member has to answer to the following question D_1 = *is the adoption of this innovation convenient for the company of the member* ?. If the answer is NO that member decides not to adopt, but the resistance to innovation will be reduced after the evalutation. If the answer is YES that member has to answer to a second question, such as D_2 = *are the environment constrains low enough to facilitate the adoption of this innovation ?* If the answer is NO that member decided not to adopt with these constrains, but perhaps later with different constrains. In case the answer is YES that member has to answer to a third question, such as D_3 = *does the member believe that the adoption of this innovation will bring some negative results?*. If the answer is NO that member finally decides to adopt and also it will promote this adoption to other members. In case the answer is YES that member does not adopt but it will consider some doubts.

The experiments and demostrations as well as the innovation agents may facilitate the process of evaluation and adoption.

We may summarize the main conditions to get an efficient *Technology Transfer Process*, as follows:

- the user must select the new technology, which has to be used to solve one important problem of the user;
- the user must choose a new technology which must be really available in his environment;
- the user must develop a strong and concrete interest for the new technology;
- the user must have sufficient resources (financial and professional ones) to acquire and to use the new technologies;

- the user must have concrete possibilities to use with profit the new technologies;
- the user must have a sufficient level of technical skills in order to manage the use of the new technology;
- the user must be ready to put in discussion his previous powers and privilegies.

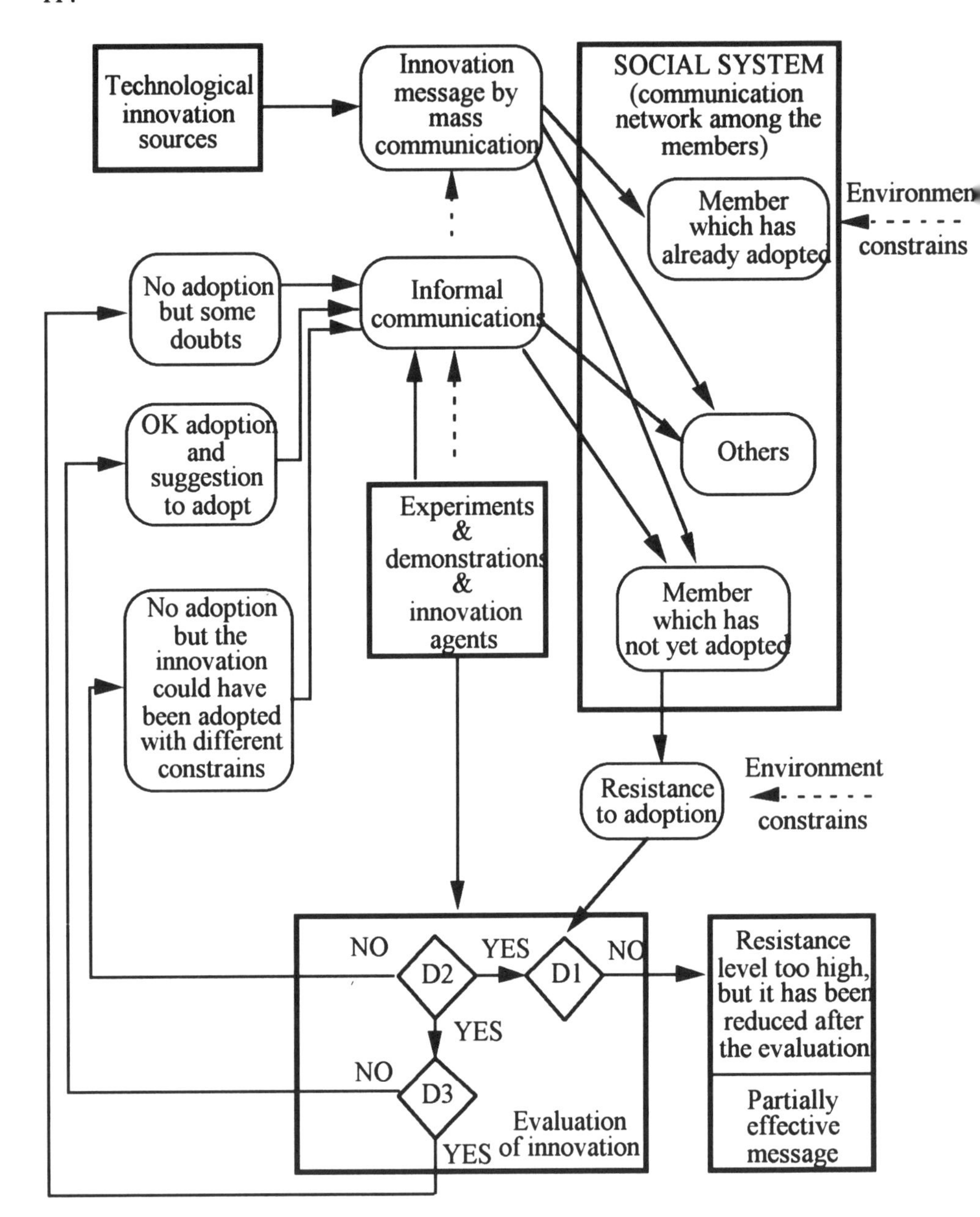

Fig. 10 A model of the technological innovation adoption

3. SCIENCE & TECHNOLOGY PARKS

A modern tool to facilitate the diffusion, transfer, acquisition and application of technologies is the Science & Technology Park. This complex system provides several services to its companies and organizations, such as:

delivering of:
  - technological information,
  - market & products information,
  - finance information,
  - etc.
- strategic & operative consultancy and managerial education,
- creation and developments of small enterprices,
- applied scientific research,
- transform of scientific results into usefull technologies,
- transfering new technologies to applications.

The Science & Technology Park is a complex system which may be considered a powerful innovative territorial system, as schown in Fig. 11, where the innovative projects help to create linkages between the organizations which offer technologies and the organizations which express innovation needs.

The innovative projects put together the achieving organizations and the financial sources, both public and private ones.

The Science & Technology Park has the following General Objectives:

- development of the economic processes,
- development of the competitive capacity of sets of companies,
- generation of new companies,
- attractiveness of the territory.

The role of the Science & Technology Park has been reported in Fig. 12 , with a special enphasis on the Technoloy Transfer Processes, Generation of new companies and Services Offering.

In particular the Science & Technology Park will promote the cooperation between Universities and Research Centres with companies and organizations, which would like to use the applicable technologies. Besides the Park will promote and develop the generation of new companies, while it will organize and offer consulting services to companies.

In doing all the above mentioned activities the Park will make innovations, as shown in Fig. 12. In the figure on the side of the services demand, I have indicated the single members of the social system with a letter (A, B, C, D) meaning that each member belongs to one of the four cells of Fig. 9.

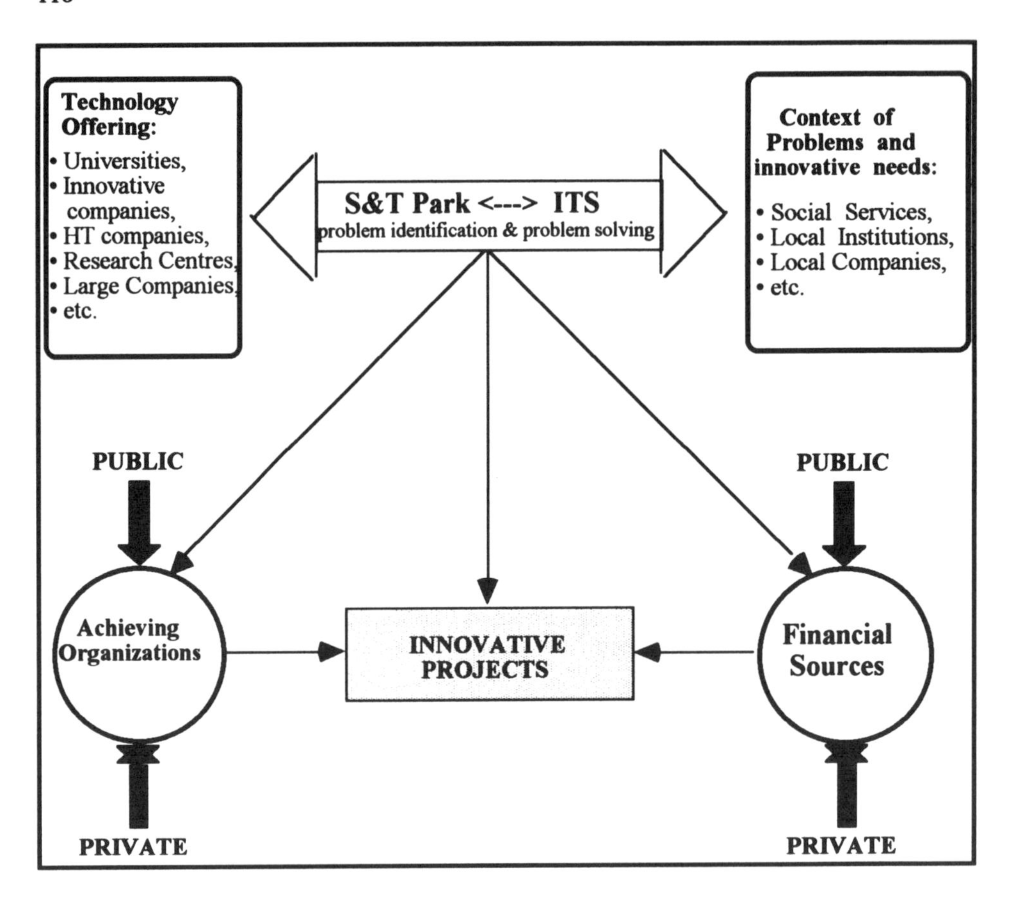

Fig. 11 (A. Romano - 1994) Science & Technology Park as a Innovative Territorial System (ITS)

In the next paragraph I will describe my personal experience made in the last four years to promote, to design and to manage the Baltic International Technopark (BIT) of St. Petersburgh, Russia, heading an Italian-Russian working group.

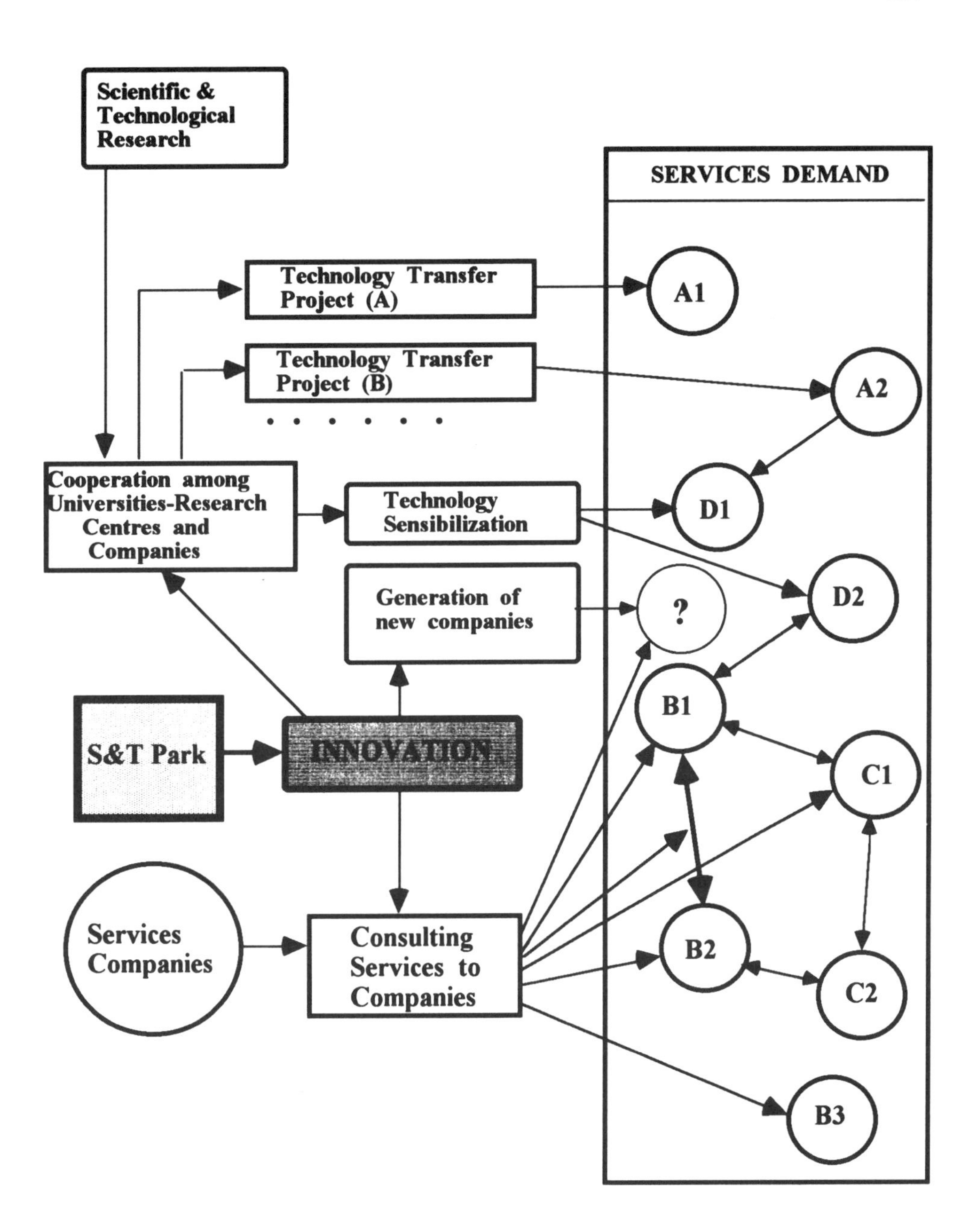

Fig. 12 The Technology Transfer Process and the role of the Science & Technology Park.

4. MISSION & STRATEGIES OF THE BALTIC INTERNATIONAL TECHNOPARK (BIT) OF ST. PETERSBURGH, RUSSIA

The BIT is managed by Baltic International Technopark (BIT) Inc. of St. Petersburgh, whose the President is Prof. Juri P. Savelyev, Rector of the Baltic State Technical University of St. Petersburgh. The BIT Inc. has been constituted on April 16 1993, by means of the transfer of the right of use of some buildings to the BIT Inc. from the partners

The BIT will Realize the Following Centers:

- Technology Transfer Center
- Innovation Center
- Certification Center
- Information Center
- International Management School

which will be included into the so called **"Science & Technology Area".**

Let me describe briefly the above structures:

4.1 Technology Transfer Center

This centre will organize the following services:

- consultancy to Russian companies about:
  - ◊ marketing and development of new products / new processes and services;
  - ◊ organization and management;
  - ◊ technology management (technological innovation, quality improvement, new technology acquisition, etc.),
  - ◊ etc.
- cooperation between universities / research centers and industries (i.e. to transfer scientific results to the applications);
- education of managers of single Russian companies;
- experimental laboratories;
- demonstration laboratories;
- etc.

4.2 Innovation Center

This Centre will provide the following services:

- monitoring new business opportunities;
- search for new enterpreneurs;
- selection of the potential successful opportunities,
- BICs:
  - ◊ start-up of small companies;
  - ◊ growing small companies;
- etc.

4.3 Certification Center

This Centre will organize and offer the following services:
- definition of standards of products / services on the basis of the international market;
- certification of products / services;
- certification of enterprices;
- etc.

4.4 Information Center

This Centre will organize and offer the following services:
- links with international data banks;
- local data banks about markets, products, companies, technologies, etc.
- international observatory on the economic and social trends of the Russian market;
- etc.

4.5 International Management School

This School will organize the following activities:
- residential short courses for international top-managers;
- master of business administration, centered on the technology management Issues;
- etc.

The "S&T Area" has the following characteristics:
- very high costs:
  - ◊ to restore structures and infrastructures;
  - ◊ to organize activities and to install equipments;
- very low income from the local market;
- very poor financements from the local and national governments;
- high possibility to get western financements for specific projects (for instant, TACIS)
- low probability that a HT industrial western company will invest to restore buildings;
- quite high probability that a HT industrial western will invest to organize activities and to install HT equipments.

From the above characteristics it follows that first of all the "S&T Area" had to indentify suitable sources to get financements in order to restore the useful buildings for the S&T Area activities. Second the S&T Area had to identify the suitable strategy to attract High-Tech industrial western companies to invest into activities and equipments. In November 1993 the Board of Directors of BIT Inc. decided consequently to organize two different businesses, as follows:

4.5.1.1 Business A:

which includes those businesses which are profitable in a short period of time. A consistent part of the profit from those businesses, must be paid to restore the buildings, where the office and the laboratories of the "S&T Area" will be located.

Possible Investors are to be identified among the sectors of Restoring Buildings, and Managing Restored Buildings.

4.5.1.2 Business B:

which includes those businesses which are profitable in a long time, i.e. the "S&T Area" activities, including also the services to the Russian companies located in cell A of the Fig. 9. The Buildings are supposed to be restored and then they can be rented. These businesses include activities in supporting the local industrial sectors, by international cooperation to develop new products/services, to penetrate the local market.

Possible investors are to be identified among the industrial sectors of HT manufacturing companies, such as Information Technologies, Automation Systems, etc..

In Fig. 13 I have reported the model of the strategy decided by the Board of Directors of BIT Inc. Where it is shown that the profit gained from the commercial businesses will be transformed into financements to be added to the local financements in order to restore the buildings where the office and the laboratories of the "S&T Area" will be located.

Under the "S&T Area" we may find the technical Centres, see Fig. 13, and each of them may be managed by a new company, with mixed capital (western and Russian).

To attract High-Tech western companies to become partners of the industrial of new companies, is necessary:

- to offer suitable and well restaured renting spaces (see the real estate business);
- to offer confortable short and long term residences (see the real estate business);
- to offer interesting market opportunities through:
  - ◊ well trained russian people (i.e. technical management of equipments),
  - ◊ well organized services, such as:
    - * information on markets, products, distribution of goods, work market, etc.
    - * fiscal & legal assistance,
    - * easy travels.
- to offer interesting technical and scientific collaborations with similar industrial Russian companies;
- to offer concrete possibilities to constitute joint ventures into specific industrial sectors with a market with a suitable dimensions.

Notice that only the first two points are related to the real estate business, therefore that last business is not enough to attract High-Tech western companies, but a development of a new advanced industrial culture is necessary.

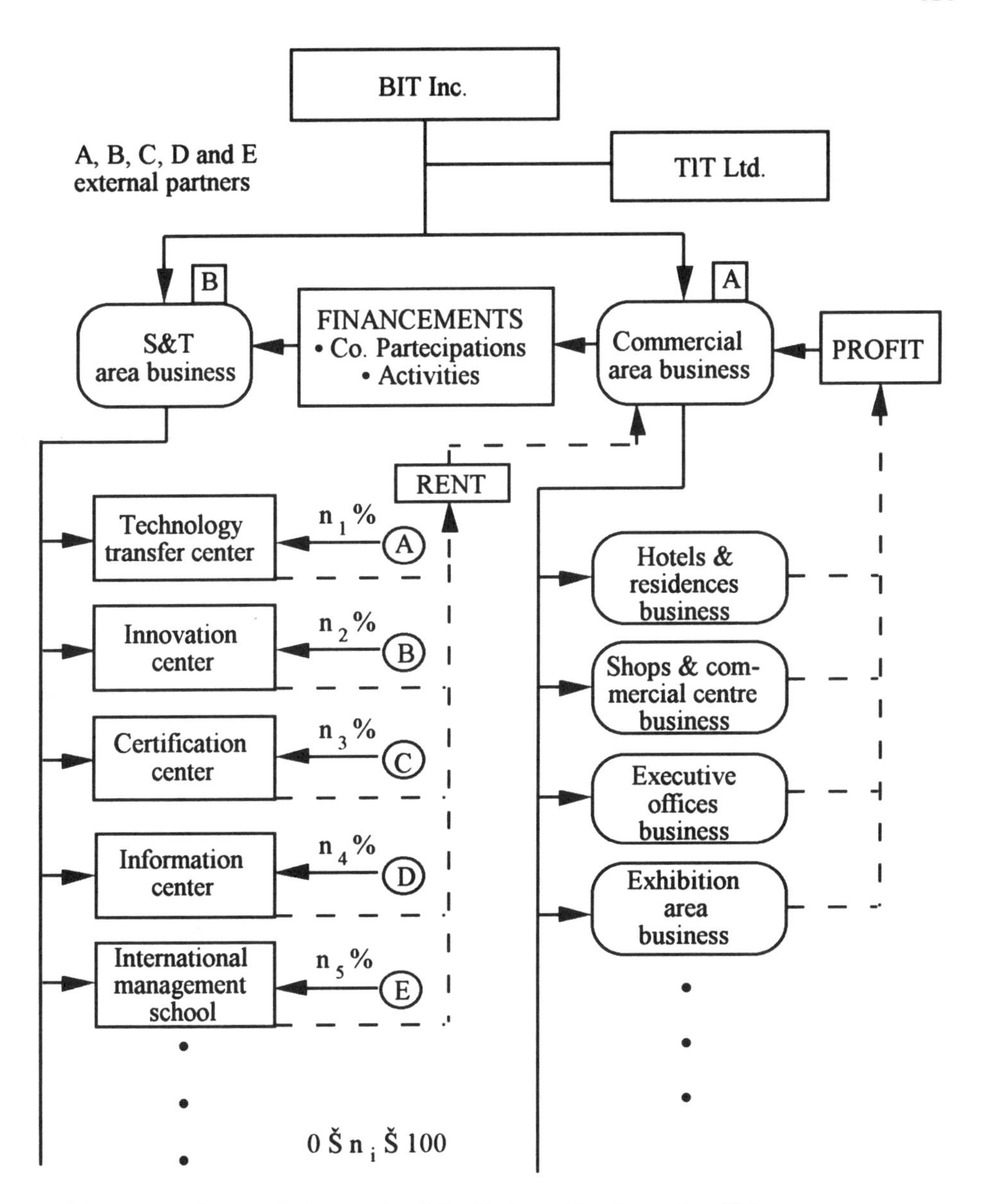

- if $n_i = 0$, for each i —> the “S&T Area Business” will be managed **only** by BIT Inc.

- if $n_i = 100$, for each i —> the only business of BIT Inc. will be the real estate one.

Fig. 13 Business structure of bit

To create this new industrial culture is necessary:

- to train the Russian managers of the BIT structures by:
    - ◊ collaboration with western experts in managing advanced innovative companies, learning to use the appropriate advanced management techniques, such as:
    - ◊ planning techniques of Innovative Projects,
        - * management of companies' network,
        - * controll of the working Innovative activities,
        - * production and logistic management,
        - * correct administrative and financial mangement,
        - * international marketing tecniques.
    - ◊ accumulation of work experience in view of the international standards,
    - ◊ investing some time of the Russian Managers in one of the western countries (stages), in order to absorb the international culture;
- to Start new Russian companies based on the international standards.

The General Strategies of BIT Inc. has been articulated as follows:

- to create links with the international world since its start;
- to create links with the Russian industrial world and services world;
- to create links with the local and national public authorities;
- to create and to develop new large international businesses in the St. Petersburgh region and more generally in All-Russia;
- to attract western investments from either banks or manufactoring and services companies, in order to increase the values of the transfered assetts, such as buildings, lands and know-how.

and therefore the mission of BIT Inc. is the following: to develop and to manage its real estate assetts in order to get money to be invested in international technological activities in order to increase the values of the buildings, to increase the future return of the investments and to organize suitable services to the local Russian companies.

Even if the two business opportunities, reported in Fig. 13, are so different, it is necessary that these business opportunities will be managed by the same managers, that means that the "real Estate Business" and the "HT Industrial business" must be developed in a tight cooperation, as shown in Fig. 14.

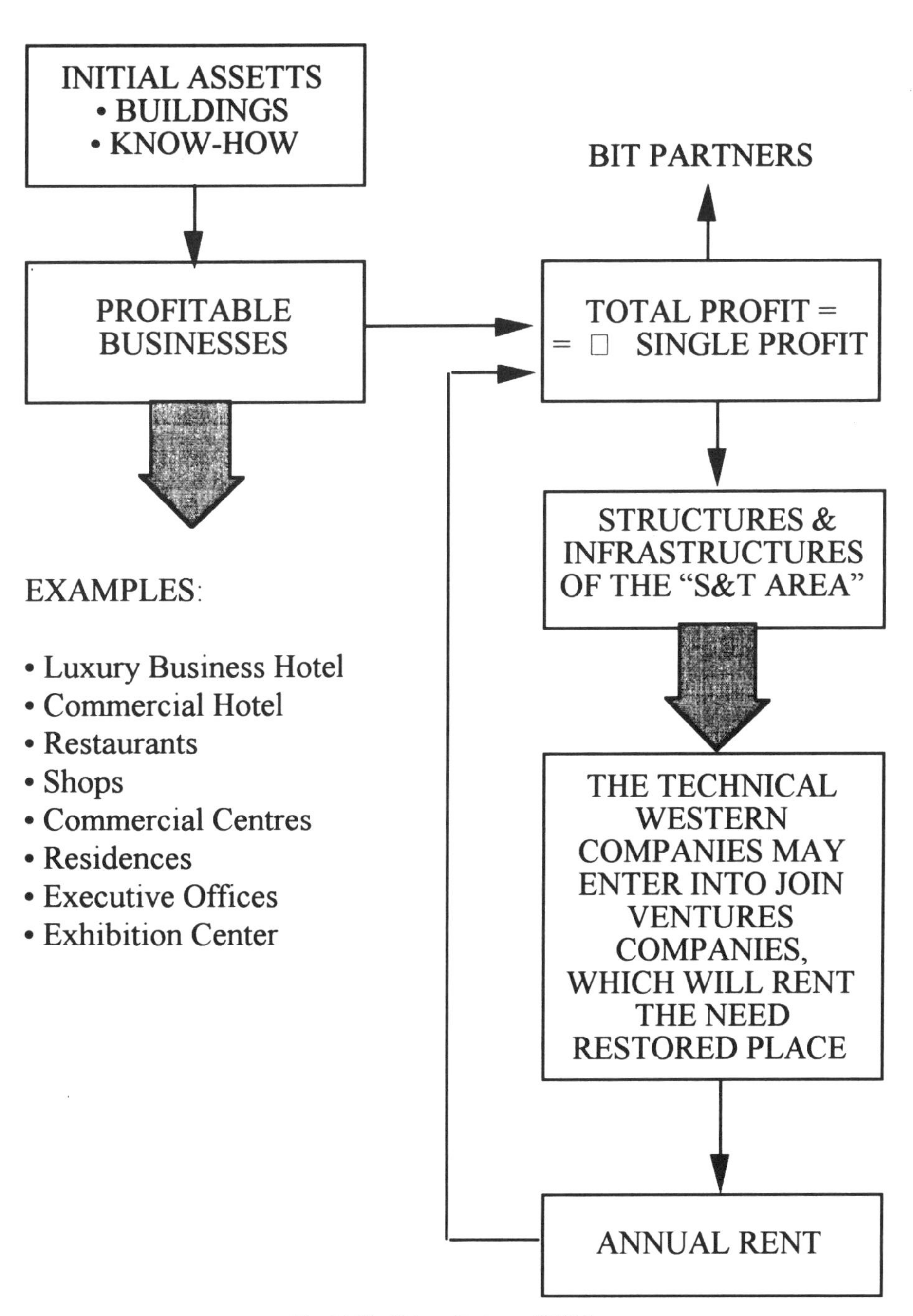

Fig. 14 The Unique Strategy of BIT Inc.

SCIENCE AND TECHNOLOGY STATISTICS IN RUSSIA:

TRANSFORMATION IN LINE WITH THE INTERNATIONAL STANDARDS

LEVAN MINDELI
Director of the Centre for Science Research and Statistics of Moscow, Moscow, Russia

1. INTRODUCTION

For many years, science and technology (S&T) statistics in Russia has been developed under the influence of the Soviet bureaucratic system of centralized S&T planning and funding. The official role of statistics has been limited to the information support of the governmental bodies. Statistics was based on gross indicators and badly suited for analytical studies. The interest in methodological investigations have been reduced, sufficient achievements of domestic and foreign statistics have been underestimated. The Soviet R&D statistical data were as a rule incompatible with the international standards because of differences in the objects of surveying, methods of accounting, data collection and processing.

Under the conditions of transition from the centralized planning economy to the market one S&T policy cannot be re-organized without a comprehensive evaluation of input and output of the national S&T system. That is why radical revision of statistics becomes a crucial component of the S&T system transformation, the background for the implementation of a new S&T policy.

2. KEY PLAYERS IN THE RUSSIAN S&T STATISTICS

In the past S&T statistics was concentrated mostly in the former USSR State Committee on Statistics. The Committee was an agency of general economic profile, and the system of R&D data collection was organized like in the other branches of the national economy (industry, agriculture, construction, etc.). It supposed universal surveying of all enterprises and, as a result, lack of the flexibility of statistics.

To clarify the current status of S&T statistics in Russia it is necessary to consider some significant factors hampering transformation of the statistical system on the base of the international standards. The major of them are as follows:

C. Corsi (ed.), Science and Innovation as Strategic Tools for Industrial and Economic Growth, 125-134.

1. The absence of the national S&T policy (in the commonly used sense) under the centralized economy. In whole it has come to the control over the fulfilment of the state plans on S&T. As a result, the statistical agency oriented toward the information demand of the central government has not been interested in deepenir and detailing S&T statistics beyond such utilitarian tasks.
   Because of the strong monopolization of the statistical services the former USSR State Committee on S&T, in its turn, was not engaged in statistics being only the user of the data provided by the Committee on Statistics.
1. Autarchy policy in the S&T sphere which had been conducted by the former USSR for decades served as an additional reason for the state statistical services not to be interested in the international standards implementation.
2. Till now remaining differences between general principles of accounting and statistics in Russia and those used internationally. This makes it difficult, sometimes impossible, to implement internationally recognized concepts, classifications and definitions into S&T statistics in isolation from the overall revision of economic and social statistics.
3. The order of the statistical data collection and processing existed in the former USSR did not promote the dissemination of ideas on the international standardization of S&T statistics. Only the central statistical agency was responsible for the all-Union data aggregation and completing the questionnaires of international organizations. The republican statistical bodies, e.g. the Russian one, have been delivered from the necessity of being acquainted with the international statistical standards.

The disintegration of the USSR has affected the statistical framework of S&T data collection. The USSR State Committee on Statistics which was in charge of statistical methodology and of R&D surveying has been abolished. The new statistical agency established on its base - the Statistical Committee of the Commonwealth of Independent States (CIS) - has not any legal rights for primary data collection. Its authority includes only methodological recommendations for the CIS countries and receiving data from their statistical agencies. This agency has very limited resources, and a large part of methodological works on R&D statistics for this body is being performed by the CSRS.

As regards the State Committee on Statistics of the Russian Federation, its role in the past has been only to collect data in Russia and to pass it to the former USSR central statistical agency for further processing. This body has not been involved in the studies on the methodology of R&D statistics and has not enough experience and resources for the implementation of the respective international standards.

The transition to the market economy has caused complication of economic and social processes, formation of a new state regulation mechanism. It obviously has required demonopolization of the national statistical services. Concrete executive bodies - ministries and governmental agencies - are to be entitled to keep statistical data collection regarding areas of their responsibility.

Unfortunately, the official statistics does not react to the modern realities in S&T and, as a result, does not meet the modern requirements of S&T policy-making. In this situation methodological and organizational alienation of the governmental statistics on

S&T from the system of forming and implementing S&T and innovation policies has become a crucial factor hampering further improvement of S&T statistics in Russia.
In this connection in 1992 the Ministry of Science and Technological Policy (MSTP) of the Russian Federation approved the Conception of Science and Technology Statistics Formation. It was proposed that the Ministry would take the authority for the development of S&T statistics in addition to those related to government R&D funding. The positive experience, for example, of the ministries of science and technology in Germany, France and in other developed countries confirms the advisability of direct leadership of R&D statistics by those governmental bodies.
The Statute of the Ministry of Science and Technological Policy, approved by the Government of the Russian Federation on July 12, 1993, proclaims that the Ministry is generally responsible for the development of the methodology of S&T statistics, implementation of the respective surveys, introduction of the international standards.
In December 1993, MSTP and the State Committee for Statistics of Russia issued a common statement aimed at joint efforts to raise efficiency of R&D and innovation statistics. According to the statement the Centre of Science Research and Statistics (CSRS) established in early 1991 and subordinated both to MSTP and the Russian Academy of Sciences is directly responsible for S&T statistics methodology, implementation of the international statistical standards, data analysis, software for data processing. The CSRS has been also authorized to represent the Ministry in relations with the interested international organizations in the field of S&T statistics.

3. CURRENT SURVEYING PRACTICE

Till 1989 the predominant concept of R&D data collection was the coverage of quite definite types of institutions (depending on the objectives of concrete surveys) versus activity-based concept recommended by the international statistics. The primary data on R&D regarded mostly the sector "Science and Scientific Services" in accordance with the so called All-Union Classification of Branches of the National Economy which was oriented to the centralized planning system and uncompatible with ISIC.
Current statistics on S&T in Russia traditionally cover several major groups of indicators:

I. R&D input.
- A. Personnel.
- B. S&T education.
- C. Fixed assets.
- D. Funding.

II. R&D output.
- A. Inventions and patenting.
- B. Prototypes of new machines and equipment.

III. Innovations.
- A. Utilization of inventions and new prototypes of machines and equipment.
- B. Production of new products.
- C. Expenditures on innovation.

Peculiarities of the existing indicators and of the available data are reviewed below.
Data on R&D in Russia (versus those of the former USSR) are in general available since 1989 when the new national survey was launched. Since than some national categories, like R&D personnel by occupation, types of R&D et al., are broadly comparable with those of the Frascati Manual.
Most of the data on R&D and innovations usually are obtained from mandatory reports of R&D institutions and industrial enterprises.
Four major national surveys should be considered.

3.1 The annual national survey of s&t activity

The survey covers all establishments performing R&D and S&T services. The data are grouped by ministries and other bodies according to the subordination of R&D units, regions, types of institutions. The survey provides full coverage of R&D activity, concordance of the methodologies of personnel and funding data estimation.
The survey reflects:

1. Indicators of R&D financing measured in terms of the total value of projects (including profit obtained during the year) and actual expenditure, both extramural and intramural (i.e. performed within the unit).
   The data are available for R&D by type of activity including basic and applied research, development (design, projects, creation of prototypes, etc.), and, besides, for S&T services (S&T information, patenting and licensing services, etc.). The definitions used are close to those of the Frascati Manual.
   The total values of S&T activity and of all projects performed by R&D units are also reported.
   In accordance with the official Instruction on Planning, Accounting and Calculation of S&T Production Cost, used in Russia, expenditure on R&D as opposed to the Frascati standards includes depreciation for building, plant and equipment, and other fixed assets intended for R&D activity. At the time this survey does not reflect breakdown of R&D expenditure by type of expenditure. Only data on the labour costs for R&D personnel are available from this source of information.
2. R&D personnel (by gender, qualification, occupation, field of S&T, region, type of institution), including researchers, technicians and supporting staff.
   The peculiarities of personnel recruitment and R&D labour organization in Russia are taken into account in the definitions used. Thus, researchers are to be graduates of higher education institutes, as a rule, with 4-5-year training (equal at least to ISCED level 6). R&D specialists with special secondary education, i.e. graduates of specialized secondary education establishments (technical colleges) with 3-4-year training (equal to ISCED levels 3 and 5), usually work as technicians. The national survey provides also separate estimation of higher education teaching staff working as part-time researchers.
   Classification of R&D personnel by formal qualification is based on categories mentioned above and connected with the Russian educational system. It proposes breakdown of personnel into groups of staff with scientific degrees (doctors and

candidates of science), higher or special secondary education diplomas.
Classification of researchers by field of S&T corresponds to the Nomenclature of Scientific Occupations, approved by the former USSR State Committee on S&T in 1988. It includes 21 fields of S&T which incorporate more than 600 detailed specialities. In whole they can be grouped into major fields of S&T stipulated by the Frascati Manual.
The indicator of R&D personnel is much more close to international standards. A significant distinction here is that R&D personnel is expressed as head counts but not in full-time equivalence. That approach has not been used in the Russian statistics, and its introduction will need additional efforts.
The survey of this kind was intended to be an intermediate stage on the way from the old statistics on R&D to the modern one, internationally comparable. Unfortunately, the further progress has been stopped by the official statistical bodies, and such a half-and-half methodology is still in use.

3.2 .The annual survey of post-graduate courses

The data are responded by research and higher education institutes with post-graduate courses, including candidate's and doctor's ones.
The survey covers post-graduate students enrolment, freshmen and output by gender, field of S&T and attendance status (full-time, part-time and distance education). The number of those defended their dissertations is also responded.
The education data does not correspond neither to employment statistics, nor to the data on R&D personnel. Data on the first destination of graduates are not available.

3.3 The survey of R&D fixed assets

Data on fixed assets in use in R&D institutions belong to traditional S&T indicators in Russia but are not commonly used for R&D analysis in OECD countries, at least not for international comparisons.
For the first time a large-scale survey of R&D fixed assets was held in 1989 and covered 5.4 thousand research and design units, higher education institutes located in all the former USSR countries. The major indicators concerned the stock and composition of the fixed assets intended for R&D, e.g. equipment.
The qualitative indicators of expensive R&D equipment referred to the age and the technical level (respectively above, equal to or less than the best world achievements). The condition of buildings intended for R&D and availability of experimental bases were also considered.

3.4 The annual survey of the development of prototypes of new equipment and instruments and their utilization

According to the existing definitions, new equipment, which has been developed for the first time in the country and is significantly different from that produced before by the

principle of operation and functional destination, is considered as a new prototype. Prototypes are distributed by industries, types of equipment, duration of development. They are also distinguished by the technical level, i.e. in comparison with the best national and foreign analogues by their technical characteristics.
Utilization of prototypes is measured by the number of those accepted for production, including absolutely new, significantly modernized or slightly modified products. Statistical reports contain also data on prototypes utilized which are based on licenses or protected by patents.
Data are available for industries, regions, types of equipment and instruments.
Thus, classifications as well as many of S&T indicators are still rather specific for Russia. Some of the indicators can be considered as supplementary to those used internationally. Among them there are indicators of R&D fixed assets (by type, age, qualitative characteristics), definitions of detailed types of development, indicators of total (beyond R&D) employment and activities of R&D units, of new prototypes as measures of R&D output, etc. This experience could be interesting for other countries.

4. TOWARDS THE INTERNATIONAL STANDARDS

Goals of S&T policy directly determine objectives of the revision of R&D and innovation statistics. From this viewpoint several main principles will be underlaid in the background of the projected S&T statistical system:

1. Provision of the realistic picture of S&T input and output. Statistics should react to the priorities of S&T policy, not only reflecting existing trends, but also allowing to foresee future changes.
2. Coverage of all relevant elements and determining factors of S&T.
3. Introduction of the international standards into the Russian S&T statistics. It should be comparable with international statistics in order to include information on S&T in Russia into international data series.
4. Concordance between S&T statistics and other ones (labour, education, enterprise, foreign trade, etc.).
5. Flexible system of data collection based on its different forms adequate to concrete statistical tasks. This requiers high-quality planning and coordination of censuses, regular reporting, sample surveys, sociological interrogations.
6. Public relations in S&T statistics: maintaining active relations with data suppliers and users, public availability of data. It includes also implementation of a program of various statistical and analytical publications on S&T indicators.

As it is supposed, at the first stage the revised system of R&D surveys will consist of the following major studies being developed by the CSRS:

- annual national R&D survey meeting international standards (since 1995);
- annual survey of government R&D funding (since 1994);
- annual survey of R&D performance by small enterprises (since 1994);
- quarterly reporting of R&D institutions on principal indicators (since the 2nd quarter of 1994) for current decision making.

The main efforts to introduce international standards are connected with the program of the new national R&D survey. The development of a specially tailored questionnaire for eventual use in collecting key R&D input and output data takes a significant place in the current methodological activity of the CSRS. It is being designed in accordance with the OECD standards, and its composition is determined by the necessity to meet, at least, the following main demands:

- to reflect specific features of R&D in transition (institutional changes, diversification of activities of R&D units, variety of sources of funds and types of property, increasing part-time employment in R&D, etc.);
- to meet peculiarities of R&D and education system in Russia;
- to provide information in the structure of the OECD questionnaires on R&D;
- to have a form appropriate for different types of institutions and statistical services in Russia.

The questionnaire was officially approved both by MSTP and the State Committee on Statistics, and includes a number of indicators which in whole can be subdivided into three groups:

1. Traditional issues for the OECD statistics as well as for the Russian one (R&D personnel and expenditure).
2. Issues which has not been developed earlier in the former Soviet statistics, e.g. distribution of R&D expenditure by socio-economic objective.
3. Issues which are traditional for statistical studies in Russia, but not in the OECD area (R&D fixed assets).

The questionnaire contains the following sections:

1. R&D personnel. It proposes head-count estimation of full-time R&D personnel by qualification, occupation, field of S&T. Measuring full-time equivalent of part-time personnel is also envisaged.
2. R&D expenditure. It is considered by type of expenditure (depreciation excluded), type of activity (basic and applied research, development), source of funds (own funds of R&D units and external funds - government, non-budget funds, business enterprises, universities, general university funds, private non-profit institutions, foreign sources). The classification of intramural R&D expenditure by socio-economic objective is also foreseen.
3. R&D fixed assets. This section is devoted to the measurement of the stock of R&D fixed assets, e.g. that of equipment.

Essential definitions and classifications (including sectoral one) used are based mostly on the Frascati Manual. The survey is oriented to using the new Russian Classification of Economic Activities and Products which is fully compatible with ISIC and NACE and has been introduced since July 1994. Besides, some additional classifications reflecting important features of the Russian economy and statistics are used, for example, classification of institutions by type of property (federal, provincial, municipal, private individual, collective, foreign, etc.).

The first annual internationally styled survey will be implemented in 1995 and is considered as a base for the further development of the Russian S&T statistics.

Government R&D funding being the major source of maintaining the national R&D potential will be another focal subject for statistical surveying. In order to obtain detailed data, in 1992 the CSRS developed the questionnaire for the survey of government R&D funding conducted by MSTP. It was aimed at collecting data on budget R&D expenditure in 1991 and appropriations (ex-ante), both actual (for 1992) and expected (for 1993), from departments, governmental agencies and other bodies (associations, academies, independent research centres, etc.). The questionnaire included indicators reflecting the distribution of budget current expenditure by type of costs (labour, equipment, others). R&D financed from own funds of R&D units and through contracts were also estimated.

Following needs of policy-makers in relevant statistical data the CSRS has developed recently a new questionnaire on appropriations for R&D from the republican budget*. Several principal requirements were taken into consideration:

1. To meet the current practice of R&D budget planning and analysis. The annual procedure of R&D budget planning includes accounting of actual expenditure of the previous year, development of preliminary plan and its final adjustment, as well as estimation of outlays required for the next year. Besides, capital R&D investment are planned in the framework of republican (federal) budget separately from current outlays.
2. To provide information for detailed comprehensive analysis of budget R&D funding. This supposes obtaining data on budget R&D expenditure by type of costs, type of activity, field of science and technology, socio-economic objective.
3. To be in line with the general revision of concepts, definitions and classifications of R&D statistics in Russia in accordance with the international standards. Concordance between statistics of budget R&D funding and a new national R&D survey developed by the CSRS on the base of the Frascati Manual is vital.
4. To reflect national peculiarities of R&D management, accounting and statistics in Russia. It is important to combine both compatibility with NABS and Frascati recommendations, and specific elements of national classifications, e.g. of socio-economic objectives and types of costs. These classifications should allow to obtain internationally comparable data regrouping detailed items.

The questionnaire includes three sections:

1. Outlays for R&D by type of costs (both actual and planned for current and next years). Breakdown of expenditure by type of costs meets usual articles of expenditure accepted in the R&D budget planning in Russia. They can be also grouped in larger elements recommended by the Frascati Manual.
2. Actual current expenditure on R&D by type of activity and field of S&T. The definitions used are in line with the international standards.
3. Current expenditure on R&D by socio-economic objective and field of S&T. The classification of socio-economic objectives is based on the NABS and at the same time reflects national traditions. This influences grouping of objectives in major groups and their disaggregation into detailed ones.

---

[1]* Improvement of the budget R&D funding statistics is being developed in the framework of the Project on R&D and Innovation Indicators in the Russian Federation being performed by Eurostat and CSRS under the TACIS Programme.

The survey is being implemented by the CSRS under the auspices of MSTP. It covers all governmental agencies receiving funds from republican budget intended for R&D.
Along with the transformation of R&D statistics, the CSRS has started developing innovation statistics as a brand new dimension of the Russian S&T statistics. Gradual introduction of market mechanisms has caused needs in new methodological approaches for comprehensive studies of innovation processes, including types and sources of innovation, stimulating and hampering factors, respective resources and output.
Medium-term activities in this area are devoted to development and implementation of an innovation survey compatible with the Community Innovation Survey (CIS) undertaken by Eurostat. As the first stage, the so called introductory survey developed by the CSRS is planned for 1995. It will be aimed at obtaining internationally comparable statistical information on innovation in industry and will cover approximately 60 thousand enterprises, including small ones and joint ventures. The questionnaire consists of two sections:

1. Main economic indicators (output, sales, exports, employment).
2. Innovation activity expressed as numbers of new products or processes introduced, and those of acquired technologies (in the form of patents, licenses, industrial prototypes, useful models, contracts for R&D, know-how, etc.).

Respondents will be asked whether they plan to develop or introduce new or improved products (processes) during next years.
It is expected that the survey of this kind will allow not only to overview a general picture of the innovation activity in the Russian industry, but also to design a population of innovative enterprises for the next stages. This will be represented by a sample innovation survey, covering only innovative enterprises, for detailed analysis of trends in innovation and determining factors.
In the long term, innovation statistics can be expanded to the service sector. Special attention will be devoted to innovation infrastructure surveys (namely those of innovation intermediary firms; information, financial, credit and legal services) and ad hoc studies of potential investors (banks, investment funds, venture companies et al.).
Other CSRS methodological activities currently cover:

- Human resources in S&T, including a new model of data collection and single surveys (stock and flows of personnel in the Russian Academy of Sciences, post-graduate training, money income of scientists and engineers, etc.).
- R&D output, including a methodology of the technological balance of payments, surveys of licenses and new technologies.

The CSRS has also an ambitious program of statistical publications. It includes reports, yearbooks, analytical reviews, thematic essays as well as methodological editions (see Bibliography). Most of them are published both in Russian and in English.

5. BIBLIOGRAPHY OF THE MAJOR CSRS PUBLICATIONS ON S&T STATISTICS

1. Science and Technology in Russia at a Glance: 1993. - 64 pp. (in Russian and in

English).
2. Science and Technology in Russia: 1993. - 240 pp. (in Russian and in English).
3. Research and Development in Russia Outlook. - 468 pp. (in Russian and being translated into English).
4. Emigration of Scientists: Problems, Real Estimations. - 48 pp. (in Russian and in English).
5. Science and Technology Indicators in the CIS. - 408 pp. (in Russian and in English).
6. Science and Technology in Russia: 1991. - 166 pp. (in Russian and in English).
7. Research and Development in the USSR, Data Book: 1990. - 60 pp. (in Russian and in English).
8. Science and Technology in the Former USSR: Analysis and Statistics. - 296 pp. (in Russian and being translated into English).
9. Directory of R&D Institutions of Russia. - 287 pp. (in Russian and being translated into English).

Forthcoming
1. Science and Technology in Russia: 1994.
2. Science and Technology Indicators in the Regions of Russia.
3. Higher Education in Russia.

FOSTERING A RUSSIAN HIGH TECHNOLOGY INDUSTRY IN AN OPEN MARKET ENVIRONMENT

ALEX COLETTI
Director of the High Technology Program, Italian Trade Commission, New York Harrison, New York, USA

1. ABSTRACT

The programs that can be developed to assist Russian scientists to find employment in Small and Medium Enterprises (SME's) are discussed as the major trust of a national Science and Technology (S&T) policy. Effective S&T policies have to fulfill real needs and have to define specific commercial goals. Whether scientists will lead their own hi-tech enterprise or work for companies, they will find highly advantageous learning software tools for modern manufacturing (CAD/CAM/CAE), and basic skills in management and in financial planning.

2. INTRODUCTION

The issue of effectively using the vast intellectual resources of Russia to trigger industrial and economic growth requires the definition of a precise Science and Technology policy (S&T).

S&T goals must be implemented with long term programs and sustained by suitable infrastructures that can ease the path from research and development to commercialization and diffusion of technology. The process is often referred to as traveling the "technology pipeline."

Large investments were made in the past on Russian S&T, and now human and technical resources, valuable for fostering the growth of Russian hi-tech SME's, can be redirected to avoid letting them disperse during the present phase of economic reforms.

Scientists have strong technical background, multi-year experience in finding practical solutions to problems, and superior attitudes to learning new skills. Therefore, it seams sensible to envision programs dedicated to effectively helping them to gain leading roles in commercial, managerial and production activities at SME's while abandoning obsolete government research programs. Structures and services could be created where scientists are assisted in the process of transferring their skills from research to design and production of hi-tech goods.

SME's need good management and careful financial planning. In order to maintain

C. Corsi (ed.), Science and Innovation as Strategic Tools for Industrial and Economic Growth, 135-141.

competitiveness they need constant technological upgrades. Small and medium enterprises (SME's) are the necessary premise for healthy economies where a variety of productions and sophisticated technical solutions maintain high commercial profits even during any possible downturn of economic cycles.
The discussion on the characteristics of the technology pipeline will help to identify programs necessary for the growth of hi-tech SME's. It will also demonstrate how Russian scientists need a basic knowledge and understanding of the skills necessary for managing the principal SME's functions competitively. Of course, the first is new product development and design with modern CAD/CAM/CAE tools, then management and finance. Scientists well equipped with elements of these three basic disciplines will be able to bring new products into production, and to gather the human and financial resources necessary for marketing products internationally.

3. OBJECTIVES OF S&T PROGRAMS

The first step in the definition of a Science and Technology (S&T) policy starts with the definition of the major societal goals to which science and technology contribute (Enabling the future linking S&T to societal goals). The goals relevant for a sustainable and competitive growth can be listed as:

- economic growth - Economic growth has important social implications, and is necessary for maintaining a good purchasing power in the domestic markets
- employment and work-force training - training is essential for incremental improvement of productions and for sustaining mobility and flexibility of the work-force
- international competitiveness - ultimately, to be successful, domestic products must be able to compete internationally
- modernized communications and transportation - product development and competitive marketing thrive in the presence of efficient services and infrastructures
- international cooperation and action - no single nation can find within itself the resources for developing new products, for maintaining competitive manufacturing, and for sustained purchasing power

The achievement of each of these S&T policy goals requires the implementation of clear programs and the support of the social forces involved. Therefore, the programs aimed at speeding the transfer of innovative ideas into production, will have to be selective according to a list of technical sectors chosen for their potentials for serving these goals.
The European Economic Community, Japan, and the United States have identified the technologies that they consider more relevant for an effective S&T policy. The list of the technologies chosen by the United States is reported in Table I (Report of the National Critical Technologies Panel).
The Table also emphasizes the different approaches that must be maintained in S&T planning depending on the goals each ministry or public agency has to fulfill. For example we can say that the list of "commerce emerging technologies" is the necessary guideline for defining the list of the hi-tech products that have to be included in the

statistics for production and import/export.

4. THE TECHNOLOGY PIPELINE

The need for scientists to acquire skills in management and financial planning emerges clearly when discussing the generic diagram that public awareness of a new technology can be expected to follow as function of time. Such a diagram can be used to represent the life-cycle of a technology leaving a government laboratory achieves commercial success while evolving through the so called "technology pipeline".

The Fig. 1 reproduces the general diagram that is often used to illustrate the development of technologies. The diagram shows the relationship between the stage of development of a technology and the amount of publicity given to the technology, which in turn affects the available investment opportunities.

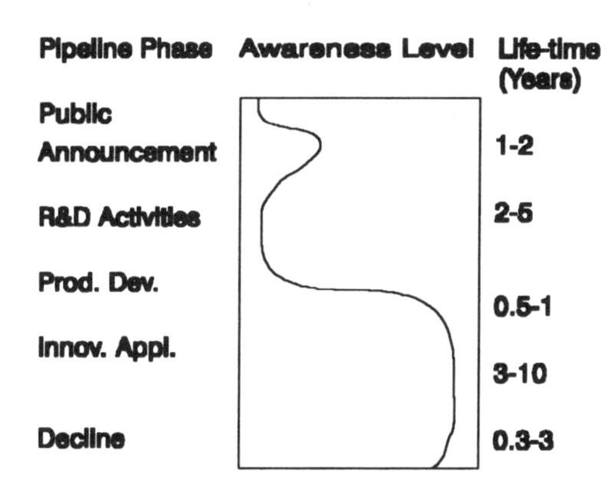

Fig. 1

Typical Diagram showing "Technology Pipeline", the levels of "Public Awareness", and "Life-cycles" for a new technology

During the phases of public announcement and R&D, government funding is important. However for a technology to move form these phases to the phases of product development and of innovative application, commercial enterprise must pick up the funding. The initial phases, describe when a new promising technology is discovered and when press conferences, government agency involvement is solicited and new product applications are envisioned.

NATIONAL CRITICAL TECHNOLOGIES

Materials	**Manufacturing**
• Materials synthesis and processing • Electronic and photonic materials • Ceramics • Composites • High-performance metals and alloys	• Flexible computer integrated manufacturing • Intelligent processing equipment • Micro-and nanofabrication • Systems management technologies
Information and communications	**Biotechnology and life sciences**
• Software	• Applied molecular biology

• Microelectronics and optoelectronics • High-performance computing and networking • High-definition imaging and displays • Sensors and signal processing • Data storage and peripherals • Computer simulation and modeling	• Medical technology
Aeronautics and surface transportation • Aeronautics • Surface transportation technologies	**Energy and environment** • Energy technologies • Pollution minimization, remediation, and waste management

COMMERCE EMERGING TECHNOLOGIES

Materials • Advanced materials • Advanced semiconductor devices • Superconductors • Advanced materials	**Manufacturing** • Flexible computer integrated • manufacturing • Artificial intelligence
Information and communications • High-performance computing • Advanced semiconductor devices • Optoelectronics • High-performance computing • Digital imaging • Sensor technology • High-density data storage • High-performance computing	**Biotechnology and life sciences** • Biotechnology • Medical devices and diagnostics
Aeronautics and surface transportation	**Energy and environment**

DEFENSE CRITICAL TECHNOLOGIES

Materials • Composite materials • Semiconductor materials and • microelectronic circuits • Superconductors • Composite materials	**Manufacturing** • Machine intelligence and robotics
Information and communications • Software productivity • Semiconductor materials and • microelectronic circuits • Photonics • Parallel computer architectures • Data fusion • Signal processing • Passive sensors • Sensitive radars • Machine intelligence and robotics • Simulation and modeling • Computational fluid dynamics	**Biotechnology and life sciences** • Biotechnology materials and processes
Aeronautics and surface transportation • Air-breathing propulsion	**Energy and environment**

Tab. 1 Comparison of National Critical Technologies with Department of Commerce

Others with no national critical Technologies counterparts.

R&D and product development phases are those where realistic assessment on feasibility and applicability of the new technology to the development of new product or products are performed

If the technology successfully survives the initial phases, it means that breakthrough products are developed which undergoes market testing. Depending on the success of the product being sold, the new technology may follow a curve similar to the one reported in figure 1.

The figure underscores how any given technology ready to leave a government laboratory, will increasingly depend on private funding for a period that varies between 2 and 5 years. During this time, the scientists that developed the technology, will have to contribute to the development of the new products along with other professionals. Only those scientists that are able to understand the value of the contributions of these other professionals in the efforts of financing, product development and marketing, will be able to maintain a leading role in the enterprise.

5. S&T POLICIES, PROGRAMS AND INFRASTRUCTURES

Programs and infrastructures must be planned to implement specific S&T policies.

Government S&T policies can be broadly classified according to their primary focus (global competitiveness):

- government investment to support R&D in existing firms and industries (war-wagon type)
- government investment in path-breaking and pre-competitive technologies to promote new industries and increase productivity of existing new product generation process (show-me type)
- government investment in programs and technologies that speed up diffusion of new technologies (uneasy rider type)
- mission oriented technology funding defence, space, health energy, belief in long term spinoffs to commercial sector (invisible hand type)

  The major programs being implemented in the United States, in the EEC and in Japan to improve the flow of the "technology pipeline" (making things better: competing in manufacturing) are the followings:
- Programs for Facilitating Innovation

At each stage of the initial stages of the technological life-cycle (research, development, and demonstration) as well as during the later commercialization and diffusion phases, the government facilitates gradual shifts of technology.

These programs may include the creation of infrastructures or programs like technology extension services and financial aid for modernizing manufacturing. Other programs might include agencies for commercialization of technology from national laboratories and universities-industry collaboration, and tapping into foreign technology initiatives.

5.1 Encouraging New Approaches

Through a variety of innovative regulatory and fiscal policies, a government can

stimulate the development of technologies and work to expand their diffusion
These programs may include programs to enforce laws on innovation and intellectual property, introduction of trade and custom duties, technical standards, environmental regulations, health, safety, ethical standards.

5.2 Learning and Work-force Training

They may include programs to support cooperative networks for small manufacturing firms, and creation of manufacturing technology centers.

6. FINAL REMARKS

The discussion of the technology pipeline and of the curve followed by the life-cycle of a new technology has shown how a "do nothing approach" is an unwise choice for modern society in the field of S&T policy. The criteria of letting free market forces determine which of the technologies developed by scientists will reach the marketplace as a product can be done only in the case of those technologies where other countries are superior or where there are special program interventions. Beside noteworthy exceptions, technology that remain excluded from S&T programs is destined to lose in the existing competition for hi-technology products.
The "natural selection process" does not assure the survival of scientists and the preservation of technical knowledge in the private sector.
It is extremely important for all scientists interested in integrating in the private sector activities to be able to access places where they can:

- acquire the necessary skills
- find basic support services (testing, computing, designing facilities)
- network with other scientists, professionals and practitioners to solve problems benefitting from each other experience on gaining access to manufacturing facilities, distribution channels and financial resources

In the United States there are several programs available to people interested in starting their own SME or in joining existing SME's. Special programs are also available for scientists interested in proposing new technologies to hi-tech companies.
The support programs for SME's are at the Federal, State and local levels. Information reguarding the programs is widely available. The various programs provide information, assistance and financial support. Universities, laboratories and research centers offer Industrial Liaison Programs, technology transfer programs and facilities open to scientists interested in developing new products.
The infrastructures dedicated to help scientists in entering the private sector should be:

- easily available
- have a good reputation
- offer assistance in the project definition and in the marketing planning phases

Different types of centers could make basic communication available (electronic data bases, newsletters, etc) as well as designing tools (CAD) that would offer the following advantages:

- teach applicable skills
- support preliminary feasibility studies
- encourage the development of realistic market assessments (cost analysis, market research, search for manufacturing facilities, etc.)

Ultimately, any reasonable S&T policy will also have to:

- create a culture for innovation in the service/manufacturing industries
- support the growth innovation through facilities commensurate to the need they must satisfy
- commensurate the efforts in transferring technologies from government laboratories to SME's on the basis of realistic production needs of the SME's and the real potential for new products to be introduced in the market by private enterprises

Since no single center will be able to effectively provide all the services that will be necessary to SME's, various private companies specialized in courses, technologies, management, financial planning and marketing in due time will be able to grow and contribute to implement the S&T policy.

7. REFERENCES

- Enabling the Future: Linking S&T to Societal Goals. A report of the Carnegie Commission on Science, Technology and Government, September 1992
- Report of the National Critical Technologies Panel, March 1991
- US Trade Investment and Technology Policies - from "Policy menus and the issue of consistency among policy makers" Robert Letovsky, Global Competitiveness, vol. 2, 1994, ISSN-1071-0736
- Making Things better: Competing in Manufacturing. Congress of the US- Office of Technology Assessment, OTA-ITE-443

R&D ROLE FOR THE IMPROVEMENT OF SMES INNOVATION, COMPETITIVENESS, AND CONTRIBUTION TO SOCIO-ECONOMIC GROWTH

LIONELLO NEGRI
Italian Research Council (CNR)
Office for Innovation Transfer, Patents, Technical Standards and Regulations (STIBNoT)
Responsible for the "Innovation Valorization and Management" Division
Rome, Italy

The transition from research to innovation diffusion - through the operational sequence which links the inception of new and/or partially innovated products, models, services, andprocesses to their put on the market and sale - is a very complex phenomenon. This complexity, emphasized by the ongoing globalization of the competitive arena, especially as regards the worlwide mounting request of quality plusses and added values in products/services and relevant processes, constrains the innovation rate in its whole, strongly penalizing the contribution of small-medium sized enterprises (SMEs) to the technological, industrial, and socio-economic growth.

Due to demand pull, industry has become qualitywise, independently from its size and technological sector. Nevertheless, most of the quality programs have priviledged manufacturing processes items, whereas a reduced attention has been usually paid to quality issues inherent the new products/services design, development, and industrialization. In particular, SMEs strengthened their quality mindedness in order to reach higher quality grades, productivity performances, and mass-production process yields, thus meeting both customers' and suppliers' requirements related to value increase and cost decrease. In fact, the analysis of the most significant initiatives runned in the past ten years with the aim of improving process quality, highlights the great relevance assigned to policies and strategies devoted to Quality Assurance, in terms of both freedom from defects/deficiencies and compliance with current standards and regulations.

Having achieved negative quality remarkable reduction, main efforts are concentrated on positive and delighting quality improvement, i. e., greater product quality abilities with reference to customer likes, preferences, tastes, and desires as well as peripheral and hidden needs.

The establishment of the customer-centered quality culture, by means of crystal clear quality objectives to be pursued through proper courses of action, leads to the

C. Corsi (ed.), Science and Innovation as Strategic Tools for Industrial and Economic Growth, 143-148.

implementation of Total Quality principles, which were effectively summarized by the Ten Benchmarks and the Four Management Fundamentals proposed by Feigenbaum[i] Special emphasis has been given to the following concepts, which were pinpointed by Feigenbaum himself:

- *quality is what the customer says it is and not what an engineer or marketeer or merchant says it is. Therefore, quality is a way of focusing the company totally on the customer;*
- *quality and innovation are mutually dependent from the inception of product, or service, development, since successful launches depend on making quality the partner of product development from the beginning, not the sweep-up-after mechanism for development problems. Quality is essential for successfull innovation for two reasons. The first is the greatly increased speed of new product development and the second is that, when a product design is likely to be manufactured in several Countries and where international suppliers must be involved very early, the entire development process must be clearly and visibly structured;*
- *quality requires continuous improvement. In fact, there's no such thing a permanent quality level, because quality is rapidly upward moving target and quality leadership demands more and more upgraded quality levels;*
- *quality is the most cost-effective, least capital-intensive route to productivity. It is necessary to change the productivity concept from the old Taylor four-letter word (MORE) into the quality leadership four-letter word (GOOD);*
- *quality is implemented with a total system connected with customers and suppliers. Technical capability isn't the principal quality problem, since differences between quality leaders and quality followers are given by quality discipline and clear quality work processes that people understand, believe in, and are a part of;*
- *a hallmark of good management is personal leadership in mobilizing the quality knowledge, skill, and positive attitudes of everyone in organization to recognize that what you do to make quality better helps to make everything better else in the organization better;*
- *quality and cost are a sum not a difference, complementary not conflicting business objectives. Good quality fundamentally leads to good resource utilization and consequently means good productivity and very low quality cost.*

From this starting point, *Quality Function Deployment*[ii] has been identified as the most effective and reliable approach to the technological development of new products/services and related manufacturing/delivering processes. In fact, it is largely experencied that to attain profitable quality the conditions below have to be accomplished:

- to build the new product, or services, quality from the beginning, being quality a company-wide process and a way of managing, which involves into a parallel development frame all departments, functions, competences, and abilities;

to manage quality and innovation as full and equal partners in developing new products, services, and processes;

- to adopt a team design approach allowing to positively integrate, up-stream as much as possible, the development and technology plans within the new products factory (product-process "marriage" by means of managerial tools such as *Concurrent Engineering, Forward Engineering, Total Industrial Engineering, Design for Manufacturing, Design Review, Process Review);*

- to find the "elegant" solution to each technical problem, applying Design Review techniques, such as *Value Engineering and Value Analysis*, searching a solution which will enhance manufacturability and serviceability with equal or better reliability, safety, and customers and suppliers combined cost[iii].

This developing way of innovation (new products, services, and processes) meets the prosumership idea[iv] and improves the relation flows within the customer-supplier system:

- prioritizing customer requisite quality, in terms both of fair price and service contents;
- tailoring new product lines with reference to the various customers classifications;
- reducing time to market and, consequently, increasing partial and total innovation rates;
- cutting down the waste of resources due to overdesign and design modifications before and/or after the new product, services, and processes are released on the field. To this aim, design and development criteria based on a system approach emphasizing parameters choice rather than tolerance statement greatly contribute to couple company and customers goals: minimize cost and selling price, the consistency being equal or better;
- increasing the company profit, as a consequence of the accomplishment between product, process, and quality control specifications. In fact, production uniformity, in terms of minimum excursions with respect to target value (also within the specification limits), dramatically decreases quality loss function since customers unsatisfactions (on the left range of the nominal value) and company additional costs (on the right range of the nominal value) are avoided[v].

A complete analysis, which considers both relationships and variables expressing the most significant quantities involved, requires to decompose the innovation and quality improvement process, thus highlighting its meaningful steps.

In this view, it is possible to draw a reference schema to quantitatively and qualitatively identify and analyze the main phases, whose input is given by the output of the up-stream. The schema[vi] focuses on the evaluation of research investment pay-off, in terms of innovation sourcing and industrial fall-out, with special reference to SMEs.

Phase 1 concerns the research activity by itself. The input indicator is provided by the financial and technical-scientific resources assigned to R&D[vii], the assessment of intangible production factors being very difficult, since they deal with brain activities, difficult to standardize, to measure, and to attribute to attained results[viii]. The output indicators concern the quantitative evaluation of research results and can be given by patents statistics, Technological Balance of Payments, data concerning products trade exchanges characterized by high technological contents, contributions to scientific and technological cooperation at the intersectorial and international level, statistical surveys on innovation dissemination, bibliometrical indexes such as the Science Citation. The choice of patent applications, as shown in the enclosed Table, arises from two basic considerations:

1. the patents represent a priviledged tool to promote technology and innovation tranfer towards SMEs;
2. the availability of data referred to different technological groups allows to use patents as input of a model which can make a quantitative assessment of innovation research "contents", also taking into account cross-fertilization phenomena. In fact, as regards the innovation diffusion pattern, cross-fertilization plays a strategic role in enhancing cooperation and partnership between large companies and SMEs (also of different Countries) and in developing the *new technology system* theorized by Freeman[ix] as a network synergically connecting scientific structures, industries, and market.

Phase 2 ("Industrialization") refers to actual transfer of the scientific research output, thus to the first application of innovation in the productive sectors - the "first successful innovation". The industrial value of innovation is tested on the field, analyzing the inherent technical problems and economic perspectives. In the proposed schema, "Industrialization" - even if evaluable from an economic point of view - means a phenomenon close to R&D, as far as difficulties in systematization and statistical treatment are concerned. "Industrialization" processes can not be measured only through inputs and outputs rating. In fact, if outputs are lowered with respect to inputs, it becomes impossible to distinguish between an effective process applied to poor-quality inputs and an inherent efficiency lack in the process itself. Effectiveness measures should be preferably global: inputs and processes should be consistent in quality grades. Furthermore, "Industrialization" opportunities being affected by diffusion networks performances and spatial distribution, it is methodologically improper to perform these measures without accurately considering the next Phase of the schema.

Phase 3 ("Diffusion of Innovation on Production") has been widely analyzed both from economists and spatial geographers, respectively focusing on implications of innovation diffusion on macro-economic variables and on spatial patterns of interaction. Phase 3 output - "Market Share of Innovation" - is a clear but hardly evaluable concept. Therefore, the explicit representation of the variable, however useful in the context of an aggregate model, should be discarded in favour of a direct connection between research and technology exploitation, as defined in the following Phase.

Finally, Phase 4 ("Utilization") furtherly develops innovation diffusion analysis, considering innovation as belonging to a "technology", whose social importance and economic weight increase on a level with diffusion. The focus on the down-stream effects related to durable goods innovation allows to draw a complete picture of this process. In this case, in fact, differences between diffusion in production and in utilization are determined both by the stock selling out and the product life. While, on the one hand, the stock selling out does not produce differences comparable in scope with those induced by dishomogeneous spatial diffusion, on the other, product life necessarily makes a substantial distinction between the production and utilization sphere. Moreover, as to partial innovation of durable goods, the utilization sphere concerns goods as a whole and does not detail technological changes introduced by innovation. In such a context, the measures of innovative changes diffusion supply an incomplete description of the consequences arising from innovation use, displaying only the temporal lags between production and market acceptance. Whereas, the utilization of the innovated good allows to monitor the innovation effects on a much wider scale. The schema generally refers to "Level of Use of the Technological Context to which Innovation Belongs", as in case of a complex direct utilization measurement an indirect quantitative analyses may be supplied by appropriate indicators[x].

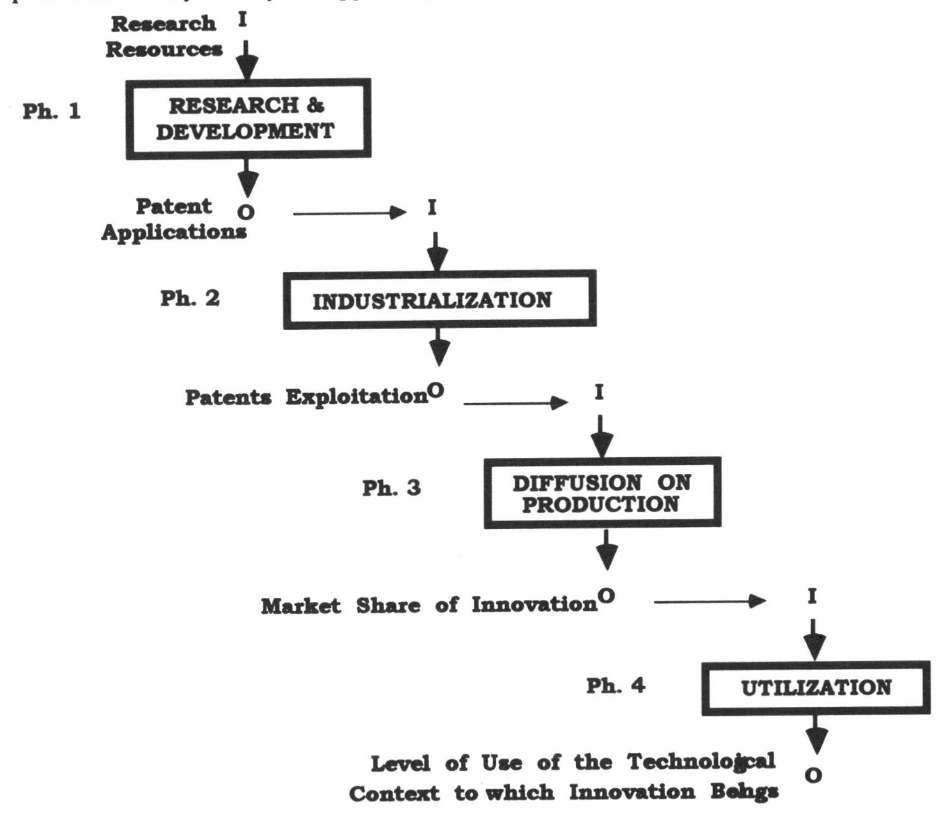

Fig. 1 The research-to-innovation flow

[ii] Feigenbaum, A. V. (1991, 3rd ed., rev.), Total Quality Control, Singapore, McGraw-Hill Book Co., pp. 828-832.
[ii] Akao, Y. (1990), Quality Function Deployment. Integrating Customer Requirements into Product Design, Cambridge, MA, Productivity Press.
[iii]Caplan, F. (1990, 2nd ed.),The Quality System - A Sourcebook for Managers and Engineers, Radnor, PA, Chilton Book Company, pp. 77-78.

[iv]Toffler, A. (1980), The Third Wave, New York, NY, William Morrow.
[v]Taguchi, G. (1983) A Introduction to Quality Engineering. Designig Quality into Products and Processes, English Translation, Dearborn, MI, Krauss International Publications and American Supplier Institute.
[vi]A preliminary draft of this schema, developed by the author in cooperation with the Department of Hydraulics, Roads, and Transportation of the University of Rome "La Sapienza", referred to the research-to-innovation chain in the automotive sector. See: BELLEI, G., L. NEGRI (1989), Scientific Research Push in Product and Process Innovation - The Case of Technological Development in Automotive Industry, 5th WCTR-World Conference on Transport Research "Transport Policy, Management & Technology towards 2001", Yokohama, Japan; S&R - Studio & Ricerca (L. Negri, Editor), No. 1, CNR-STIBNoT, Rome, Esagrafica, 1990. Text available on request.
[vii]The assigned resources can be split up into: pursued objectives, research typologies and institutions where the activity is carried out, financing sources, science/technology areas, technological sectors generating innovation, classes of products/services to which R&D is devoted, skilling and qualification of the research staff.
[viii]It is also questionable whether it is correct or not to put research under "productivity" constraints and the existing body of literature is rather limited.
[ix]FREEMAN, C. (1979), The Determinants of Innovation. Market Demand, Technology, and the Response to Social Problems, "Futures", No. 11.
[x]For example, with reference to the road transport sector, data concerning accidents and their severity can offer a proxy measure of the diffusion scale of new on-board equipments and devices to improve active and passive safety, considered as innovations belonging to the "automotive technological context".

CONSIDERATIONS ON REALIZATION OF TECHNOLOGY TRANSFER IN RUSSIA

ANDRZEJ GORAK
Dortmund University
Dortmund, Germany

1. INTRODUCTION

The main objective of the workshop was to find out the ways to enhance the technology transfer between Russian and Western countries. In my contribution I would like to describe:

- the possible areas of technology transfer,
- the potential sources of the technology,
- the problems, connected with practial realization of technology transfer,
- the ways for improving the technology transfer.

2. ADVANCED TECHNOLOGY AREAS IN RUSSIA

The Russian science was ever concentrated on some areas which had more or less direct connection to military industry. During the workshop some of these areas have been identified (e.a. microrobotics, information technology, biotechnology, space technology etc.). In my opinion one area has not been pointed out sufficiently: it is the energy sector, which is supposed to be the strategic area of Russian economy.
In Russia some economical projects for energy saving have been initiated. They are administrated by the central govenrment (for example Russian Energy Saving Fund) or by local authorities (for example Russian Demonstration Zones). Some great trusts have been grounded (for example "Kriokor", "Metran") which group up to 20 companies and up to 100 000 coworkers. These trusts originated from the previous military companies and produce now waste-heat boilers for municipal and industrial boiled-houses, electric melting complexes, reactive power controllers, electromagnetic and heat process measuring instruments. The companies carry out the works in energy audit and power consumption optimisation.
Independently on all areas, indicated by the organizers of the Workshop as potential

C. Corsi (ed.), Science and Innovation as Strategic Tools for Industrial and Economic Growth, 149-151.

collaboration fields the following areas should be also considered for future cooperation:

- measures, control and regulation of the fuel and energy comsumption,
- construction materials and insulation,
- electrical appliances and equipment,
- auditing and consulting.

3. SOURCES OF TECHNOLOGY

There are two main sources of technology in Russia:

- universities and Academy of Sciences
- military research institutes.

Almost 80% of researchers are grouped around Moscow, St. Petersburg and some regions in Syberia. The potential of Russian science lies more in sophisticated theory (which is more and more "absorbed" by western countries through brain drain) then in the matured production methods. The number of researchers at the Academy has been reduced drastically. It suffers in insufficient salary and laboratory equipment. Many research institutes can pay salary to their coworkers because they rent the buildings or equipment to private companies. Therefore, there is a very limited potential for the innovation market within Russia. The offer for technology market outside Russia can be practically only provided by previous military research complex. These are for example the institutes of process control, chemical engineering, turbine technology etc.

4. PROBLEMS OF TECHNOLOGY TRANSFER

The main difficulties in the efficient technology transfer from Russia lie in the following reasons:

tendency to make "monkey-money". The long-term cooperation is very difficult, because the Russian partners are mostly interested in the rapid profit.

- "spy-syndrom", which does not allow to break the psychological barriers during the first contacts,
- language barriers stop the information exchange (lack of knowledge of English and/or Russian),
- unstable tax policy motivates Russian partners for illegal actions,
- blockade of information dissemination by managers on the middle hierarchy level

5. WAYS FOR TRANSFER IMPROVING

Because of big potential (human and resources) in Russia it is worthwile to invest in technology transfer from/to Russia. There are some measures to improve it:

- search for young people, speaking English and with an extensive professional experience
- grounding of technology centers, supported mainly (but not exclusively) by western untries
- large information actions about the programmes of EU, directed to Russia

- cooperation with specialists, working in Europe and speaking Russian.

The idea of grounding of the common technical language (CAD/CIM/CAM) can be supported if the specialists in Russia will be educated in the use of such tools. Up to now there is a very thin layer of SMEs in Russia and therefore the conditions for the introducing of such expensive tool must be carefully examined.

INNOVATION AT BRUNEL SCIENCE PARK

PETER RUSSELL
Director of Brunel Science Park
Uxbridge, Middlesex UK

At this conference on Science and Innovation as Strategic Tools for Industrial and Economic Growth, I would like to present to you a paper of examples of innovation in the Science Park situation.

1. Working Knowledge Transfer Limited, was set up by five people who were working together in an organisation which was part of a Government Agency, when a change in Government policy resulted in their redundancy. They developed a multimedia package for Apple Computer Systems whereby they took a video output from a Management Conference and turned it into a full interactive multimedia system combining text, still images, sound and moving video. This company is currently developing an interactive multimedia system that will help managers and management students to analyse their management skills, plan their future development and build a portfolio of their competences (Appendix 1).
2. Cimio Limited, which is an expert company in CAD CAM, one of whose founders wanted to define a language that would enable computer aided design data to be assessed independently of the system creating it. This would enable a computer programmer to be able to write a code independent of any set CAD system and also to perform the data conversion, from a system already in existence, to a new system without knowing the background computer language or data structures of the first system.
   A new language has been defined by Cimio who developed a compiler for the language which will automatically generate code to perform data conversion from one model to another (Appendix 2).
3. Mr. Russell pointed out that in Russia, there are also many natural materials and told of The International Tin Research Institute at Brunel Science Park who have worked very hard in finding new uses for tin, some of which because of the need to become environmentally friendly, with new uses for solder and even caps for wine bottles as these are becoming environmentally unfriendly because of the lead content. Work in tin has also been expanded to produce a fire retardant including a compound being put in clothing to minimise fire risk. The International Tin Research Institute would like to work with Novosibirsk if arrangements can be made (Appendix 3).
4. Derivative Technology Limited who have, over ten years, had many problems in developing this company but now has developed and had tested a new compressor

C. Corsi (ed.), Science and Innovation as Strategic Tools for Industrial and Economic Growth, 153-155.

which is supported by pressurised bearings rotating on ceramics, using air as a lubricant (diagram Appendix 4a). This has great import in production of food as well as the pharmaceutical industry and a great deal of interest has already been shown. This discovery was mentioned in the London "Times" Newspaper about three weeks ago, which has resulted in an investor from Washington, trying to negotiate the production of this particular compressor with the inventor and owner of the company, Dr. Richard Gozdawa. Cash predictions on the profit from this invention are almost unbelievable.

Dr. Gozdawa also had other inventions which are of interest including an oil free air/gas compressor (appendix 4 and 4a).

5. Biocompatibles Limited who have experience that it takes a long time (12 years) to take something to the market place. Professor Dennis Chapman observed that when in a controlled environment he looked at a biological membrane as a barrier, certain components of the membrane did not allow any bloodclotting and he thought this would make an excellent coating of medical devices to stop clotting, which often causes premature death. Through his discovery that it was polymerizable phospholipid that would give this non clotting element, he has used it in many new products. The first to reach the market place was a new contact lens containing this lipid and even though it was only taken to the market place this year it has achieved sales beyond expectation. He has also applied it to catheters, heart valves, springs (a type of miniature support for coronary arteries) and also in heart bypass circuits. There are also applications in non health care, in anti fouling paint, cosmetics and moisturers (a lipid is a natural moisturiser). In addition Professor Chapman has an interest in other discoveries, one of which is an artificial blood but, although the Army and Navy were interested, they did not proceed to fund the research. His drug delivery system still needs some work to be done, therefore the challenge for Biocompatibles is having more years of research possible, if their first group of products helps to pay for the future research (Appendix 5).

Mr. Russell mentioned, in passing, that there is a United Kingdom Science Park Association which has a membership of 52 Science Parks in the UK which does include affiliated memberships and UKSPA invite any Science Park in the world to consider affiliated membership because it is a method of keeping up with developments in the United Kingdom; anyone interested in membership should contact Mrs. S. Cooke, United Kingdom Science Park Association, Aston Science Park, Love Lane, Aston Triangle, Birmingham B7 4BJ (telephone 0121 693 4850).

Mr. Russell stated that he has many ideas on the subject of international technology transfer but one that he felt could be explored, if sufficient funds can be found from NATO, is specialist workshops for SME's where probably leaders of their Science Parks would come along with companies with specialist interest. He suggested there were possibilities in (1) medical technology, (2) CAD CAM systems (3) Microelectronics and (4) work for existing materials and hopefully creating new materials, etc.

In closing, Mr. Russell said it was apparent that, in Russia, there is a great strength of sound research which has taken place over many years in the Science Academies and felt sure that some help could be given from commercial experts, possibly from the

Western Countries, if they were taken into confidence to look at and identify the discoveries that could be innovative in their own right and make a useful contribution to the economy of Russia. He had examples from his University (Brunel) that had been overlooked by the inventor and exploited by others because the innovation potential had not been realised.

COMMENTS ON WORKSHOP PAPERS

J.R.TURNER
University of Durham, UK

During the last two days I have had the privilege of hearing a range of very interesting papers on the subject in question and I would like to share with you some of my observations and experiences in science-based business activities.

To put my remarks into context, I should tell you something about my own background. First of all, I have to admit that I am not an academic. Most of my career has been spent with British and American research-based companies in the chemical industry. For sixteen years I was a profit accountable general manager, responsible for sales, production and research in a division making a range of speciality chemicals, which we sold word-wide, including Eastern Europe. I now manage the Science Park attached to Durham University and I am also the University's Industrial Liaison Officer.

The problems we are discussing here are also being experienced to varying degrees in most countries with advanced scientific communities, especially those which have concentrated a high proportion of their research on defence matters. The key problem is how to wean scientific research away from dependence on Government support and make science and scientists learn to survive and thrive in the private business sector. This problem must be solved quickly to prevent the erosion and degradation of the national science base. Frankly, I have to say that I do not believe that such a solution is possible.

The rapid reduction, world-wide, in defence related research can not be taken up by private business. The private sector will only support research which is relevant to their core businesses and from which they can expect to see a financial return within a reasonable period. In fact, many companies have reduced or even closed their in-house research departments and expect to rely on buying-in specialised technologies and new products to meet their needs in a rapidly changing market place. Government research laboratories are also being closed or privatised, with a resulting loss of long term fundamental research projects which have no place in the modern commercial organisation.

Many areas of research are declining world-wide and redundant scientists are finding new careers away from research, in industry, commerce, consultancy and activities such as teaching and even science park management. Other groups of scientists and complete government laboratories are being encouraged to retain their skills by turning themselves into private contract research companies, selling their services to specialised market sectors.

C. Corsi (ed.), Science and Innovation as Strategic Tools for Industrial and Economic Growth, 157-158.

The new commercial approach to research has now affected government support for university research in the U.K. and I believe in other countries as well. In future, academic scientists must show that their proposed research will contribute to "national wealth creation" in order to win research grants. Industry and commerce is now seen by the U.K.. Research Councils as the "customer" and the universities as the "providers".
Because of this policy, the universities are now very keen to sell their research and consultancy skills to governments and private business.
Despite the new dependence on bought-in technology by governments and big companies, there is fierce competition for their custom from an increasing number of suppliers. Organisations making a living from contract research and technology transfer will need to match their scientific skills with strong marketing and management capability if they are to survive.
Unfortunately, it is my experience that most research scientists do not make good business people. They lack the sales and marketing skills to promote and sell their products and services. More seriously, they lack the guile, business judgement and ruthlessness to survive in a competitive commercial environment. They are too trusting and too dedicated to their speciality! Given time, help and training some can learn the necessary lessons and will succeed.
In our science park at Durham, we look very carefully at scientists proposing to start up businesses. We check their business plans and their financial support, but we look even more closely at the people themselves. Businesses operated only by one or more scientists have poor performance records. Personally, I look for well balanced teams, consisting of one or more good technologists/scientists and a good, marketing/sales person with experience of the markets which they plan to enter. They should also have a source of good financial advice. Most of our start up companies meet this specification and have also benefited from training at the Durham University Business School, which specialises in the management of small and medium sized businesses.
In conclusion, I believe that there is a good future for independent research laboratories and research companies, provided they develop good "products" in the form of special skills, new technologies and innovative products or services which are relevant to the needs of their prospective clients. But, to succeed, these organisations must be fully supported by professional marketing, sales and finance departments, which can ensure that their products and services are delivered profitably, on time, to the proper quality and at the right price. Technology transfer has become a highly competitive game which only skilled professional teams should play.
Thank you for your attention ladies and gentlemen.
J.R.Turner B.Sc., Ph.D.
University of Durham, 1st November 1994

DYNAMICS OF INNOVATION

OLIVÉRIO D. D. SOARES
CETO - Centro de Ciências e Tecnologias Opticas Lab. Fisica, Fac. Ciências, Universidade do Porto
Porto, Portugal

1. ABSTRACT

Dynamic aspects of the innovation process are reviewed. The culture of innovation is emphasised while the concept of a networking environment is proposed to respond to the practices of integration of technologies in the innovative procedures.

2. INNOVATION A MULTIDIMENSIONAL PROCESS

Innovation is certainly a dynamic process with different phases having different time constant responses.
The innovative process starts with an internal phase characterised by the creative process. The feasibility of an idea is demonstrated for the case of a successful creative cycle. This internal phase is shielded from influential market procedures.
To become an innovation the invention has to be examined in terms of the industrial and market oriented potential while niches of market opportunity have to detected and evaluated[xi]. The external phase is initiated. A multitude of aspects should then be considered as for any product or service already in the market. However, as an innovation specific aspects will have to be considered: design engineering of the invention to persuade the investors and ultimately the potential users of the usefulness, safety, environment harmless, competitiveness potential, etc.
The existing standards on those undergoing adoption should be taken into consideration in terms of compliance with specification and assurance of transportability, interoperability in line with the globalizing market tendency of and delocalization of production and services networking.
The original concept of product evolution by design has matured to a broader amplitude of so called dynamical factors of competitiveness. They must be considered into the design but also to create a selective differentiation and diversity to both market acquisition and enlargement.
The lifetime of an innovation is sensitive to many external factors so that some artificial anchoring of the consumer should be used to protect from a quick ageing. Quality

C. Corsi (ed.), Science and Innovation as Strategic Tools for Industrial and Economic Growth, 159-163.

reputation is the obvious but a general acceptance of a trade mark will support the inevitable transmutations required to extend the lifetime of the newly introduced product/service in tune with consumers spectations or hiding effect from concurrence competition.

3. INNOVATION THROUGH CASE-STUDIES

Innovation is an old and young discipline. Old as a main agent of fostering the progress we all enjoy. Young as a generalised strategic tool for for economic growth. It is indeed part of the generalised concept of management of the permanent change. The abundancy of the literature has not produced yet the universal recipe for a guaranteed success.
Innovation has to be considered as a culture, an attitude, a vital ingredient all along the productive cycle, including pre and post-productive phases. Therefore, it is largely assumed that case studies are a rather valuable methodology. As in other fields some of the uncessful cases tend to be most educative to beginners.
The analysis of variety of cases will bring to evidence some cares such as the avoidance of one to attempt to reach a universal compatibility of design but rather an adoption of a common language (i.e. universality at specific layer(s) of the system). This is particular valid for areas of CAD/CAM and CIM[xii]. As a further example on may forcasts strong difficulties for software based instrumentation for the general consumers market not based on windows software environment!

4. INNOVATION FINANCIAL DYNAMICS

Funding innovation involves risks of specific nature correlated to the financial dynamics of the process, Fig.1.
Main features to be considered relate to the need of further funding exceeding the start up capital while the market has to discover and accept the novel product/service.
The need for continued innovation is also stressed, in respect, to the laws of market declining and the obvious need for a product lifetime that should at minimum recover the investment.

5. THE CULTURE OF INNOVATION

Innovation has to be perceived beyond new finished products or services but rather a state of mind. Therefore, as a culture it does not necessary mean, a new advanced technology, that occasionally the market could not feel the need as yet. Innovation is above all based on advanced production and market strategies and in the management of change at large.
As a state of mind, peopleware is vital in the innovation process. It is on their hands and mind the dimension of the success and the amplitude of percolation of innovative character throughout the entire cycle of production, distribution, maintenance and recycling.

6. IMPLICATIONS OF THE NON-DETERMINISTIC CHARACTER OF INNOVATION

Albeit the planing one does not expects a deterministic unfolding of innovation. It is related to external factors whose relevance the economical analysis can help to establish. However in the family of the concerned multiple parameters some are random behaved others unpredictable either on space, time or both. These undeterministic character could intervene in setting a safe desired rate of innovation success and eventually in setting priorities.
As part of the strategy to obtain a safeguard, priority should be given to the higher added value innovations.

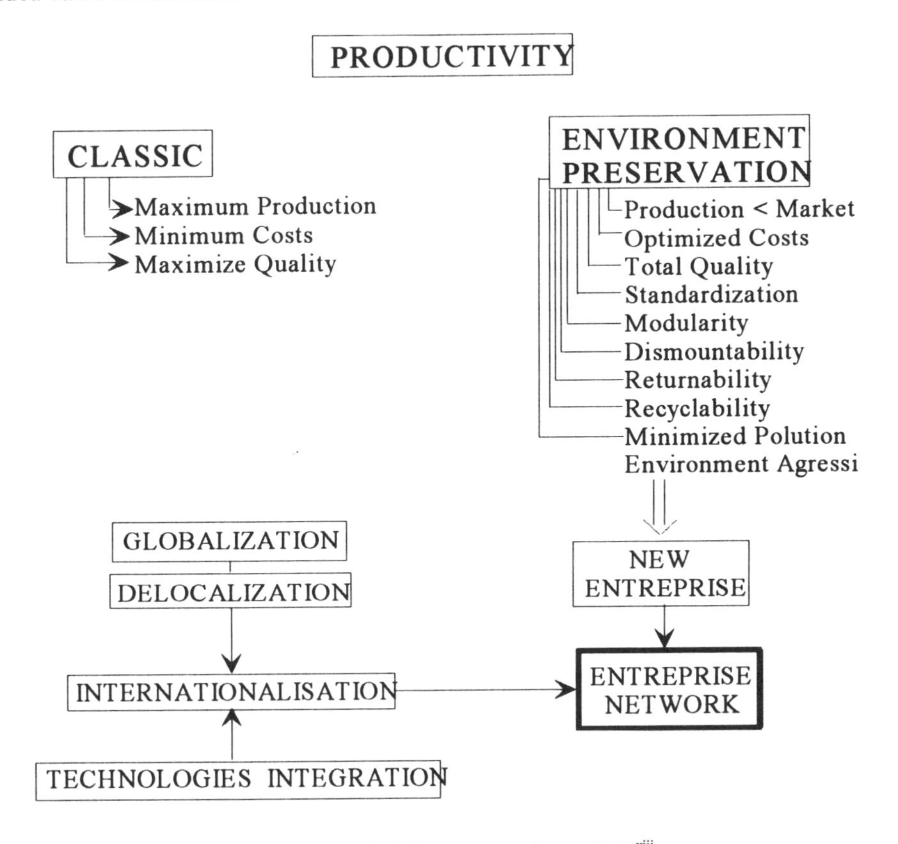

Fig. 1: The productivity concept methamorphosis[xiii].

7. INNOVATION AND SYSTEMATIC COOPERATION

The fact that robots are increasingly being introduced to manufacture high quality products proves that the success of innovation should not be necessarily linked to the

products or even the production technology.
Modern manufacturing technologies and capital mobility rather emphasises the innovation as a systemic concept. Innovation appears strongly entangle with management. The extensiveness of the requirements to bring innovation into a market success and the global character of market competitiveness seems to recommend that a fertile ground for innovation could be reached via networking[xiv]. The concept of network means here cooperation sharing mutual benefits.
Countries and regions portrait common goals because they face common needs of economical growth. No country could expect to reach the highest competitiveness in all fields. Countries and regions would have to opt to specialised domains.
Complementary, the evolution trend is for integration of technologies. Sometimes the innovative character is almost the result of an answer to a spotted need by an well elaborated integration of technologies available elsewhere.
The Airbus project, the EUREKA projects among many others strongly emphasise the value of networking and joint-venture projects.
Networking is also in line with modern relations of trade. Indeed, any country that wants to sell should be ready to buy. Further, in progressively opened market economy competition is played directly among companies. For an innovation to enter the competition game creating the knots to anchor the networking could prove to be a rather competitive advantage and result in shortening the desired response time from the market to innovative efforts by companies.

8. THE PORTUGUESE DIMENSION

In what concerns innovative cooperation with Portugal one may refer its roots as earlier as the time of the discoveries, in the XVI century. A large experience has been acummulated and is now part of the cultural heritage and continued practice.
Today a series of missions, visits and intercompanies contacts have occurred with materialised results.
Further progress can be steer up with the cooperation of IAPMEI - Instituto de Apoio às Pequenas e Médias Empresas and via the AIP-Associação Industrial Portuense, and other available mechanism.
The case studies of established cooperation (outside of the scope of present text) could also be beneficial to the fostering of a larger mutually proficous collaboration.

9. REFERENCES

1 O.D.D. Soares Economic Development of Photonics in Europe Dubrovnik, SPIE (1982)
2 C. Corsi This volume
3 O.D.D. Soares Applied Laser Tooling Martinus Nijhoff (1987), 1-24
4 A. J. Hingel A New Model of European Development and Network-led Integration FAST (1993)
5 O.D.D. Soares Novas Indústrias e Infra-Estrutura de Ciência e Tecnologia na

Transmutação Industrial e Proteção do Ambiente Didaxis, Riba de Ave (1993)

[xi] O.D.D. Soares Economic Development of Photonics in Europe Dubrovnik, SPIE (1982)
[xii] C. Corsi This volume
[xiii] O.D.D. Soares Novas Indústrias e Infra-Estrutura de Ciência e Tecnologia na Transmutação Industrial e Proteção do Ambiente Didaxis, Riba de Ave (1993)
[xiv] A. J. Hingel A New Model of European Development and Network-led Integration FAST (1993)

"OUTBOUND TRANSFERS OF TECHNOLOGY IN A COOPERATIVE PERSPECTIVE"

BORISZ SZÁNTÓ
Academic Research Unit, Technical University
Müegyetem rkp 9, T ép, 201/b
H-1111 Budapest, Hungary

1. COMMENTS ON CONTRIBUTION

My own 3 years of experience say that for the time being technology transfer from Russia to the West can be practically done only via people's informal ways of cooperation. The main problem we face in this respect is not financial but the lack of common trust: Russian inventors generally trust neither their own formal (governmental or private) organizations nor their partners; on the other hand Western partners do not trust the professional qualities of their Russian partners. As far as I can judge Russians are reluctant to teach the experience they possess to their Western partners. Their ideas may be outstandingly brilliant, the lack of mutual confidence as well as language and cultural barriers make the outcome rather vague.

2. CONTRIBUTION TO THE ROUND TABLE DISCUSSION

As one of the key-speakers I tried to show that Central and Eastern Europe lost to its Western partners just because of its failure to catch up with the pace of their innovative development. The innovative economy of industrialized countries gradually developing from the late 1950's has led into a qualitatively new stage of human development called "the era of globalizing technology". The changes in the world resulted in a long and grave crisis within the former socialist countries of Central and Eastern Europe. Drastic political changes in 1989-1990 did not stop this alarming tendency and the policy of new governments did not bring solution to the problem. The decline of innovative capability in these countries still continues. This is why the aim initiated in the title of this conference is so important for all of our countries and the integrating Europe as well.
Addressing myself to Russian participants I have to admit that the painful transformation of Russian science and its gloomy future makes their reasoning regarding innovation rather science-sided or so-called "wishful thinking". Almost all of the suggestions and plans presented during the session by our Russian colleagues were about implementation or export of scientific results and the future of scientific communities. Even the justification composed in advance for the present NATO

C. Corsi (ed.), Science and Innovation as Strategic Tools for Industrial and Economic Growth, 165-168

workshop concentrates on "the transfer of know-how to industrial application", "task of utilizing the advanced level of research", as well as on "the necessity to identifying the specific items".

As far as my opinion is concerned the export of know-how has nothing to do with the problem of innovation, since innovation is a continuous process of self-development, idea generation and management of its technological advantage and economic benefit on the market. The innovative economy does not require one's own scientific results to start the process, since the necessary know-how can be purchased. By character this is a nonlinear process, which cannot be dealt with traditional linear conveyor-like models of labour division, or the paradigm of forced implementation ("vnedrenie") of scientific results. Neither can we apply item- or product-logic to the process of innovation. The innovative self-development surely requires even more advanced know-how which again might be bought on the market, or replaced by results of our own targeted research when our capability to master the technology is high enough to keep up the pace of innovative development vis-à-vis our competitors. For the innovative entrepreneur it is like going upstairs along the ceaselessly descending escalator. Once stopped, without the governmental assistance it is very difficult to reach the advanced level of development.

We may and have to rely on the most progressive and ambitious innovative entrepreneurs. An innovative entrepreneur knows well what is his strongest side, what and how he would like to develop to become the best in his field of activity. The state choosing to rely on the strongest entrepreneurs can help them to develop themselves. This is the essence of the technology policy implemented by Japan and several other countries in this century. The government of the country ought to know what is the strongest side and market position of its industry, how it can be made even stronger, and what will be its favourite industry 10 years later. Knowing this the government should prepare a strategy of implementation based on collaboration between entrepreneurs, banks and the state institutions. The aim of technology policy is to create technological advantage in some sectors of the industry and agriculture, implement it on the world market, and to keep this advantage until the next innovative move forward is prepared. Thus in some cases in accordance with technology policy the state purchases may serve the entrepreneur, but even in that case it is the level and pace of development of entrepreneurs that really counts and not the purchased product itself.

It does not matter what kind of government institutional structure is implemented by such technology policy, if it is the pivot of the governmental activity and not just a peripheral activity of an Agency or Ministry of Research and Technology. One needs a model to be followed, selected very carefully and adjusted according the national pattern. But it is the concrete technology policy that provides real content to the model. I have to warn, that the US model is not really suitable for our countries, since its spearhead is still the military technology, with the state investing in military sector to create the required technological advantage for the whole industry. Japanese model of self-development, on the other hand, means aiming at macro-level to the more sophisticated technology, creating today the technological advantage of tomorrow and making educational and research preparations for the day after tomorrow.

Creation of technological advantage is the matter of knowledge and ambition. Motivation is an organic part of this process, indeed, but it is more self-expression and success, then just the profit that really counts. One cannot provide enough incentive to those who are not innovative to change their aptitude.
Investment in R&D or the number of scientific institutions not the precondition of innovation but one of several, and may be not the most important factor in implementation of innovative strategy. Buying licences may be more to the point, and than one is better to invest in R&D of the third or forth life cycle of one's present technology or in the next one. The R&D and licensing data are used by statistics to catch the process of development. The crucial point in innovation is not the investment in R&D, but the pace of self-development of the innovative entrepreneur.
"In-house research" also cannot be the only way to create technological advantage. In most industrialized countries it is usually the way to establish the firm's individual expertise and core knowledge. Innovative self-development and market requirements nowadays mostly mean acquisition of lateral knowledge (it can be 90% of the total knowledge required) from outside of the firm. Therefore the measure of competitive innovative capability is not the in-house R&D capacity but the ability of the management to set and solve their problems.
We consider scientific institutions the part of a country's culture and heritage. It is very painful to see this national intellectual treasure being commercialized and falling apart. On the other hand, we must realize, that intellectual heritage is connected to the level of education and as such it is the subject of national cultural policy. Technology policy is an engineering approach to the problem of socio-technical evolution of the firm and the country. Naturally it depends on the level of general culture, but islands of culturally higher industrial segments can be created, if needed. This is how less developed countries are rapidly approaching the most advanced ones. To use the existing R&D capability to modernize the production process and product technologies as general solution cannot be applied. It is always a matter of particular entrepreneurship and individual consideration.
My suggestion is to separate the question of scientific establishment and national Academy from the very rational subject of technology policy. The partnership that ought to be established for this purpose is that between Entrepreneurs, Banks and supporting Government.
Generally, I am rather pessimistic about the possibility of real reforms in our countries, if our efforts remain separated from or fragmented off of the main course of political development. One cannot improve but insignificantly any particular segment of society. Even if one succeeds partially, the rest of society and the conservatism of people will neutralize it and rapidly bring everything back into the good old order. This is what I have experienced in the 1980's in Hungary.
I am also pessimistic about the solutions of an economically and especially culturally highly developed and traditionally cooperative country being applied directly in a country with strong traditions of centralistic hierarchy. I doubt that they will work properly. We better consider the society and economy of our countries an organism being organically transformed into a modern one, and not cured from a disease. And if

this is so, the first thing we need is the ambition and wisdom on the highest level of control in the country. Without that it will always be just a new case-study of a restructuring. If the predominant course is mere stability no significant change will occur.

2.1 My recommendations:

1. To prepare a careful analysis of the present innovative capability in the country and the process of regression that led us to it.
2. To select models of advanced countries but to be careful with their application.
3. To separate the subject of science policy and the subject of technology policy, since they are really directly interconnected only on the high level of development.
4. To organize trainings of quality control and innovation management.
5. To formulate the concept of national technology policy.

INDEX

BICs 112
CAD/CAM/CIM 8
centralized S&T funding 119
centralized S&T planning 119
centre of excellence 19
certification center 113
commerce emerging technologies 130
common technological language 8
concessional taxation 49
cooperation 1
critical technologies 23
demand pull 60
Design and Engineering 15
diffusion-transfer process 93
EC 1
economic and industrial growth 5
economic reforms 55
European Technological Language 12
financial resources 54, 96
industrial sectors 93
information center 113
information flow 94
information resources 97
innovation 1, 138
innovation center 112
innovation in decline 74
innovation potential 45
innovation technology 5
innovation centres 5
innovation centres and technology transfer programmes 5
innovative activity 13
internal R&D 13
international management school 113
knowledge 5
maintenance 8
managing the innovation change 5
market-driven model 85
marketing 15, 66
material resources 96
multibusiness impact 23
national critical technologies 130
Network 5
pre-competitive technology 62
privatization 57
production investment 15
quality 138
R&D 15
R&D cooperation 9
ranking criteria 23
Russia 1
Russian cultures 34
science and technology (S&T) statistics in Russia 119
science parks 5, 87
SMEs 1, 13
technological change 69, 94
technological poles 5
technologies 1
technology driven model 84
technology filières 65
technology information 8
technology matrix 22
technology pipeline 129
technology planning 20
technology push 60
technology transfer center 112
technology transfer process 105
technology transfers 5, 6, 34
telematic link 12
Total Quality 8
trading companies 66
Training Centres Network (ITTTC) 6
transversal technologies 8
unsolicited technologies 47
value analysis 139
value engineering 139

T5-AVZ-823

The Failure of White Theology

Martin Luther King, Jr. Memorial Studies in Religion, Culture and Social Development

Mozella G. Mitchell
General Editor

Vol. 3

PETER LANG
New York • San Francisco • Bern • Baltimore
Frankfurt am Main • Berlin • Wien • Paris

Patrick Bascio

The Failure of White Theology

A Black Theological Perspective

PETER LANG
New York • San Francisco • Bern • Baltimore
Frankfurt am Main • Berlin • Wien • Paris

Library of Congress Cataloging-in-Publication Data

Bascio, Patrick.
The failure of white theology: a black theological perspective/ Patrick Bascio.
p. cm. — (Martin Luther King, Jr. memorial studies in religion, culture, and social development; vol. 3)
Includes bibliographical references and index.
1.The Failure of theology. 2. Black theology—History.
3. Afro—Americans
—Religion. I. Title. II. Series.
BT82.7.B37 1994 230'.089'96073—dc20 93-24030
ISBN 0-8204-2257-6 CIP
ISSN 1052-181X

Die Deutsche Bibliothek-CIP-Einheitsaufnahme

Bascio, Patrick:
The failure of white theology: a black theological perspective/ Patrick Bascio.
- New York; San Francisco; Bern; Baltimore; Frankfurt am Main; Berlin; Wien; Paris: Lang, 1994
(Martin Luther King, Jr. memorial studies in religion, culture, and social development; Vol. 3)
ISBN 0-8204-2257-6
NE: GT

Cover design by Geraldine Spellissy.

The paper in this book meets the guidelines for permanence and durability of the Committee on Production Guidelines for Book Longevity of the Council on Library Resources.

Printed in the United States of America.

DEDICATION

To Kendrick Radix, Maurice Bishop (assassinated) and Unison Whiteman (assassinated), founders of the Grenada Revolution and dearest friends.

ACKNOWLEDGMENTS

I owe a debt of loving thanks to:

– the late Prime Minister of Grenada, Mr. Maurice Bishop, for the influence our intimate friendship had in the development of my theological thinking.

– Black Theologians James Cone and Cornel West in whose classes and shared conversations I matured in my understanding of the importance of the black preacher and the indigenous black church.

– Mr. Kendrick Radix, former Attorney General of Maurice Bishop's Revolutionary Government of Grenada, who continues to educate me in my academic and social enquiries into the living dynamism of the black diaspora.

– the black communities of Africa, the Caribbean and New York City, in which I was privileged to work.

– the librarians of Harvard's Widener and Divinity School Libraries, and Ms. Gayle Pershouse of the Episcopalian Theological Library, in Cambridge, for their kind assistance in my research.

– Ms. Gloria Dugan, of Beaver Island, Michigan, for the technical and other support provided during the writing of this book.

– Liz McBride, of Newport, R.I.; Lennie and Laurel Earnshaw, of New Hampshire; Norma Blaize, Grenada Representative to the Organization of American States; and the Jesuit Community of Hawthorne Street, in Cambridge, Mass.

– Margie and Mike Koczat (aunt and uncle) of Methuen, Mass. for the good pasta!

CONTENTS

Introduction 1

Chapter One: **Ideological And Local Theology** 13

Chapter Two: **Slave Religion And The Dream Of Abolition** 39

Chapter Three: **The Sources Of Black Theology** 67

Chapter Four: **Major Black Theologians** 97

Chapter Five: **A Critique Of Black Theology** 123

Chapter Six: **Problems To Be Dealt With** 133

Conclusion 159

Introduction

Black theology, as it has evolved in America, has its roots in ante-bellum slavery and the religious practices that developed within the slave system. Its development continued during the post-bellum period, as Emancipation allowed blacks to found their own churches and to discard the religious facades acquired during slavery. Although Emancipation did not confer the full rights of citizenship to the black community, the opportunity for blacks to practice a religion based on cultural and social customs rooted in their African past expanded. Black theology considers itself a liberation theology, a theology in which political and social ideology is expressed and socio-political strategies are planned, in the ongoing struggle for the full realization of civil rights and social acceptance in a predominantly white society. It is a local theology, for its roots are in the black community, from Africa, through slavery to modern ghetto America. It is a theology that comes out of a church with a long history of political activism, a church in which religion and politics are considered two sides of the same coin. In an effort to represent the black community, to make its cry for understanding and brotherhood heard in the white community, black theology has engaged in an impressive critique of classical theology. This critique, if found to be valid, would add new insights to our understanding of the Christian message, and its theological analysis.

Since black theology has aroused the conscience of America, especially on the questions of racism and civil rights, a study of its arguments is in keeping with America's boast of having established a humane pluralism which is a shining example to the world. If the black community has become the "conscience" of America, then black theology is of interest to every American. It is not surprising that what is black has become a source of wisdom for what is white and strong, for "God chose what is weak in the world to shame the strong." (I Cor. 1:27.)

Black theological themes are of vital importance to white America's psyche, for, without them, the possibility of the white community living an authentic existence in American society is remote. Because the privileged position of the white community is inevitably related to the suppression of civil rights in the black community, every white person born in this country is caught in a web of his/her ancestors' making, from which there appears, as yet, no way to exit. Given even the most sincere desire to avoid participation in racism, nothing in the American political system enables the white person to project an effective protest, one that radically energizes society to abandon its self-inflicted curse. There is left, only, the option of crying in the wilderness, the option of "betraying" one's own kind. There is no easy solution. Historically, many whites have tried to cope with the burden of racism by participating in black/white coalitions, but even that avenue has been fraught with difficulties. The abolitionist movement of the nineteenth century, led by middle-class whites, proved to be so paternalistic that black activists, like Frederick Douglas and his allies, were forced to withdraw from black/white cooperation for lack of participation in the decision-making process. W. E. B. Dubois dealt with white paternalism, as did Booker T. Washington.

The first break with this approach was the black mass movement organized by the Jamaican, Marcus Garvey. His influence led to "separate" black organizations like the NAACP and the Urban League. With Martin Luther King Jr. came the all-black coalition, with sympathetic whites this time not involved in leadership roles. Nevertheless, the black/white coalition, during particular historical periods, did contribute toward a creeping progress, laying the foundation of the Emancipation Proclamation and other milestones in the black man's search for equality. The coalition is not dead, but it does need to be re-awakened. The Black Congressional Caucus motto strikes a practical, if chilling, note: "We have no permanent friends in America and no permanent enemies—only permanent interests." This is far from satisfactory from the perspective of Christianity; it may be all the black community can work with at this time.

Is There a "Black" Religion?

The subject itself begs the question: is there a "black religion" sufficiently distinct to justify a separate category, terminology and theology? Is black religion merely a more animated form of white Christianity? The questions imply that black religion may be a matter of external detail, not a matter of substance; that black religion is nothing more than a cryptonym for the prevailing white man's expression of the Christian faith. Such an evaluation would not give serious consideration to the possibility that the American practice of Christianity derives from a European ethic and cultural history, one that has little in common with the Afro-American experience.

It was not until the Black Theology Project of 1977 (The Black Church and Black Community: Unity and Education for Action) that white America was even aware that the black community was engaged in serious theological exploration. The white academic community was taken aback at the theme of the Project, the indivisibility of salvation and historical liberation. The historically personal-salvation-focus of the white theological community left it intellectually and religiously incapable of responding in any articulate manner. Theology was, perhaps for the first time in American history, discussing serious social issues. The realization that a mature theology was developing in America outside the white theological establishment caused both curiosity and anxiety. Could there possibly be a "black" theology? It took years to acknowledge that the theology practiced in America had always been "white."

Organizational work for the Project began in the Spring of 1976 under the direction of Rev. Muhammed Kenyatta, and Sister M. Shawn Copeland, a Dominican nun. It attracted such participants as black theologian Sister Jaime Phelps, of the Chicago Catholic Theological Union, folk-singer Frederick Douglas-Kirkpatrick and prominent black theologians James Cone and Gayraud Wilmore. It major themes were:

- Jesus is the Black Messiah, the liberator.
- Failing to distinguish between salvific and historic liberation, white religion has contributed to black oppression.
- The black church reflects its African past by sacralizing the world.
- The black church preaches the theme that black is beautiful, an important element in the development of self-esteem and self-awareness.
- The black church is the institutional framework for the black socio-economic agenda.
- The black church will expose the negative impact of capitalism on minorities in America.

Black Religion Not Jeffersonian

Black religion resisted the intellectual, logos Christianity handed down to generations of Americans by Jefferson, in the philosophical form of Modernism. This intellectual religious anthropology constructed a God who is inapproachable, transcending even artistic attempts to reveal His image and personality. Black religion participated in a post-modernist attempt to reverse this particular God-image, by defining knowledge in terms of the capacity to live in constant touch with the world around us. Black theology humanized the Greek logos concretely and historically by grasping, intuitively, God-Incarnate, thus taking the lead in the postmodernistic concept of religion as the integrating component of man's historical journey. It reflects its pragmatic roots in African religions, focusing on everyday concerns, in contrast with the Christian European obsession with personal salvation. By refusing to divide the world into secular and sacred, Black theology nurtures and preaches a humanistic wholeness comfortable with a personal, familial God. Robert Hood observes:

> Theologically, the Afro traditionalists offering the supreme god as one who reigns over a cosmos that includes not only humankind, but also spirits, divinities, ancestors, and other forces, inanimate and animate beings.[1]

By refusing to focus on individual salvation, black religion breeds a social cohesiveness that is the envy of the rest of God's children. In the midst of, and in spite of, a white Christianity that justified slavery and racism, it called its members to live a simple life of authentic existence, together.

Black Community Has Its Own Needs

Since it is only in recent years that scholars have elaborated on the role of culture in religious practice, it may not be appreciated that the concerns of white churches may not necessarily coincide with the critical religious needs of black people. There has always been a notable absence of the black agenda when leaders of American Christian denominations meet to discuss religion, so the black church became the focal-point of developing amongst blacks a self-consciousness proper to the black psyche, one that

---

1. See Robert Hood, *Must God remain Greek? Afro Culture and God-Talk* (Fortress Press, Minneapolis,) 1990.

reflected African religious concern for social and political expressions. The assumption that white religious expression should fulfill the needs of the black community was based on classical theology's failure to understand the critical role of culture in religious expression. Prompted and nudged by black theological writings, the white church is only now coming to grips with this reality, a long-delayed acknowledgment that the neglect of culture was but one facet of the racism lodged deep within its theological structure.

Black theological analysis has, where appropriate, negated the white Christian's interpretation of the gospel message, substituting in its place an interpretation relating to black social existence that has, since it is biblically-based, more universal application. "We shall overcome," a slogan used today by many white religious and social organizations, is but one example. Cornel West observes that the black historical experience and the biblical texts used in the black church form a symbiotic relationship, "each illuminating the other."

Black theology is, in part, the response of black men of religious feeling and commitment to what they perceive to be the failure of the theology of the dominant culture in American society to understand and relate to their spiritual needs. Further, black theology is a protest against the continued reality of racism amongst white Christians in our society. Since black theology is a response to theological failure, and a protest against social conditions, it is sometimes perceived as a negative theology. I shall look carefully and with sympathy at black theological writings, unearthing the very positive messages communicated to both the white and black communities.

An Ancillary Purpose

An ancillary purpose is to examine the reasons for the establishment of a "separate" theology, its contribution to the religious life of the black community, and its role in raising the social consciousness of white America. As one can expect, there are similarities and differences between classical and black theology. The similarities are not surprising, since most black theologians were introduced to theology and trained in that discipline by seminaries and professors of the classical school. My purpose will be to focus on the differences between the two disciplines, since their differences are far more important to this study than are the similarities.

Where Was the White Church?

One might expect that the resources that were available to the American theologians associated with the mainline churches, resources which include major Universities such as Harvard and Yale, would have produced a social Gospel that did not exclude the very group of Americans most in need of its saving grace. John C. Bennet, of Union Theological Seminary, New York City, agrees with those who charge that classical theologians neglected the problem of black oppression.[2] Even more shocking to the Christian conscience is the insight of the German philosopher David Friedreich Strauss, who wrote at a time when slavery had only recently become illegal. He maintained that the abolition of slavery was not a result of humanitarian/religious efforts on the part of the Christian church, but rather the result of the intellectual analysis done by the much maligned Enlightenment. Human rights, he claims, is a philosophical rather than Christian concept.[3]

The fact that the Christian churches of the colonialist period were unable to inform the European-American conscience with a respect for the dignity of human beings who just happened to be black is a phenomenon worthy of theological and, ultimately, social analysis. That this theological analysis did not take place until representatives of the victimized race acquired theological expertise, speaks to the continued lack of interest among white Christian theologians in a theology that addresses itself to the great moral questions related to the rights of the human person. The harm done to the black psyche as a result of the "white" agenda of American Christianity is described in a book edited by Emmanuel K. Twesegye:

> Unfortunately and predictably, most of the colonized people aspired to become like their colonizers in order to become human once more. But by so doing, they negated themselves, their own civilization, values, history, religion, culture, art, language, education, technology, food and even dress...this colonial problem of miseducation and loss of identity was particularly notable amongst Africans in America.[4]

---

2. John C. Bennet, *The Radical Imperative* (Philadelphia: Westminster Press, 1975), p. 119.
3. David Frederick Strauss, *Der Alte und der neue Glaube* (Bonn: E. Strauss, 1903), p. 56.
4. *God, Race, Myth and Power, An Africanist Corrective Research Analysis*, edited by Emmanuel K. Twesegye (Peter Lang, New York), 1991. p. 9.

Meager Efforts

To the black theologian, the mild liberal reformism which has characterized the white theological community had the effect of calling for change in a status quo which violated human rights without, at the same time, calling for fundamental changes. "Tokenism" has become the word that characterizes such meager efforts. The history of Christianity teaches us that spiritual conversion has always had to stand in opposition to social impediments which boldly interject themselves into the Christian praxis, the story of slavery being the most dramatic historical example. Such impediments have continued to prevent the explosion of goodness which the Gospel message can ignite; the task of the Christian theologian is to engage in battle with these impediments. That the timing for such battle is closely related to the economic and cultural development of the "white" world is clear. The fact that it has taken nineteen hundred years for the Christian theological community to address itself seriously to the moral questions arising from the relationship between men of different racial backgrounds, is itself an indictment of the vitality of that community. The recent Jewish holocaust of the Nazi era reminds us of the need for constant vigilance in the struggle against the insertion of impediments to Christian life and practice. That the historical theological enterprise is not free from impediments which arise from time, place, culture and race was stressed by Helmut Gollwitzer, Professor of Theology at the Free University of Berlin. He observed that the phenomenon of black theology is one historical consequence of colonialism and the slave trade.[5] It is only in recent times that the sociological and theological effects of the creation of subcultures within a society have been recognized. The sociological reality that the development of a subculture brings with it the legitimation of a subordinate and superior social position certainly has significance for a Christian theology that proclaims the brotherhood of man, based on the shared Fatherhood of God. That the complexion and the physiognomy of the white man was both a sign of superiority and the human ideal had to become a subject worthy of theological analysis. However, it did not come soon enough to remove the blemish from the thoughts of great men. David Hume, the noted Scottish philosopher, speaking of the intellectual abilities of black people wrote that they were naturally inferior and had

---

5. Helmut Gollwitzer, "Why Black Theology?" *Union Seminary Quarterly Review* (Fall, 1975), p. 17.

failed to produce, anywhere, a civilized nation.[6] Thomas Jefferson in his "Notes on Virginia" states that blacks are much inferior to whites in the art of reasoning, and that in imagination they are "dull, tasteless and anomalous."[7] Suffering from the intellectual schizophrenia common to many of his Enlightenment colleagues, Jefferson did sponsor a proposed ordinance of 1784 that would have banned slavery in the new Western territories after 1800, a proposal that was defeated because of the absence of a Mr. John Beatty of New Jersey on the day of the vote. Jefferson later remarked of that defeat: "Thus we see the fate of millions unborn hanging on the tongue of one man, and heaven was silent in that awful moment."

The Dream

William McClain, a pastor of Union United Methodist Church in Boston and lecturer at Harvard Divinity School, saw black religious practice as reflecting a liturgy and theology arising from the circumstances of a people living on the cutting edge, from a classical theology more "rationalistic than radical."[8] The particular genius embodied in black religious expression did not prevent the black community in colonial America from dreaming about living in a land where all men, black and white, would live, work and worship together. This dream is described by Langston Hughes[9] and Margaret Walker.[10] Hughes wrote "America never was America to me" but did not hesitate to "swear this oath" that the American dream would one day be a reality.[11] Margaret Walker observes that "Neither the slaves' whip nor the lynchers' rope" succeeded in crushing the Black belief in the New Jerusalem.[12]

In the 19th Century, American blacks pursued a dream rooted in the Declaration of Independence and in the Constitution. They dreamt that

---

6. "On National Characteristics," quoted in Richard H. Popkin, "Hume's Racism," *The Philosophical Forum*, Vol. 9, Nos. 2-3, p. 213.
7. Notes on Virginia," quoted in Winthrop Jordan, *White Over Black: American Attitudes Toward the Negro, 1550-1812* (W. W. Norton & Co., 1968), pp. 436-437.
8. William B. McClain, "The Genius of the Black Church," *Christianity and Crisis*, Vol. xxx, No. 18, November 2, 1970, p. 250.
9. James Mercer Langston Hughes, America's most notable black poet and writer, was born in Joplin, Mo., Feb 1, 1902, and died May 22, 1967. His first poem, "The Negro Speaks of Rivers," was published in 1921. He published his autobiography *The Big Sea* in 1940.
10. Margaret Walker, eminent black poetess, was born in Birmingham, Ala. July 7, 1915. Her first volume of poems appeared in the *Yale Series of Younger Poets*.
11. Langston Hughes, "Let America be America Again," in Langston Hughes and Ana Bontemps, eds., *The Poetry of the Negro 1746-1949* (New York: Doubleday and Company, 1949), pp. 106-08.

one day the democratic ideals enshrined in these historic documents would be actualized, that blacks would one day be citizens of a loving community, where they would be invested with all the rights, privileges and responsibilities invested in white men. It was clear to the black community of the 19th Century that they were not included in the rights and privileges described in the Constitution and the Declaration of Independence. This awareness was reinforced by the institutional patterns of Church, State and neighborhood. The church rationalized its acquiescence in the accepted oppressive patterns by concentrating on preparing black souls for the heavenly kingdom, while at the same time communicating to the black community that, contrary to prevailing racial patterns, blacks were children of God, brothers and sisters of the white community. The brother/sister "relationship" remained, of course, at the mystical level of "spirituality."

It is estimated that out of one million blacks living in 19th century America, 50,000 had become Christians. It is significant that although the "New Jerusalem" dream had not yet been realized, black Christians continued to pursue it with petitions to State legislatures and to the Congress. It was in the context of these efforts at self-help and the pursuit of equality before the law that the black church was born. And it is because of the continued search for the "New Jerusalem" that the black church remains today the place where blacks worship and where they plan their socio-political strategies. It is not surprising then that the theology derived from such a church is characterized by the dual purpose of spiritual and social liberation. The founding of separate black churches was, in part, the black community's response to the failure of while politicians to extend the privileges of citizenship to America's blacks. Some black churchmen found in the Gospel a mandate for a revolutionary path to the realization of full citizenship in American society. Among the most notable of these was Henry Highland Garnet, a Presbyterian minister who preached that it was a matter of Christian obligation to revolt. A resolution passed in Bethel Church in Philadelphia in 1917 under the leadership of the Reverend Richard Allen boldly proclaimed that the descendants of slaves had a right to share in America's abundance.

---

12. Margaret Walker, "We Have Been Believers," ibid., pp. 180-81.

The Dream Lives On

The move toward a more aggressive black unity and solidarity under the rubric of the black church began in these early days in Philadelphia, Silver Bluff, South Carolina (where the first black Baptist Church was founded) and in other communities throughout the land. The modern black theology of freedom and liberation, of the renaissance of black religion, is the continued affirmation of the black community's ongoing struggle for spiritual and physical liberation. Frederick Douglas, the great black orator of the mid-1800s, declared that those who profess their allegiance to the concept of freedom, and yet criticize the kind of agitation necessary to acquire that freedom, are men who want crops without plowing up the ground.

The leaders of the early black church learned to keep a delicate balance of prayer, hope and agitation in their pursuit of freedom. The dream of the "New Jerusalem" has been deferred, but it has never been abandoned. Martin Luther King, Jr., at the march on Washington, in August of 1963, reechoed the theme when he spoke of his dream that one day this nation would rise up and live out the true meaning of its creed: "We hold these truths to be self-evident, that all men are created equal." This sermon was given in the shadows of the Lincoln Memorial and the choice of this site is fraught with historic symbolism, for after a hundred years of non-fulfillment, a great American black man returned in person and in spirit to the original place of the issuance of the Emancipation Proclamation to enunciate it again in his own language and style.

A cynical Malcolm X referred to the March on Washington as "The Farce on Washington," but perhaps a longer view can find in Martin Luther King, Jr.'s efforts seeds of hope and eventual realization. King placed himself in that long line of black preachers who, historically, have been a beacon of light and energy to his people, the one most in touch with God, assuring them that He is waiting in the wings of history to guarantee the black community safe passage for those who love Him and do His will.

The Time For Union Has Passed?

The time has probably passed for black/white organic union in religion, but not for back/white cooperation in a spirit of mutual respect. The success of such cooperation depends upon the determination of the white church to overcome the racism within its own ranks and institutional structures. The leadership of the white church can still determine how its intellectual, spiritual, and economic resources can be so reallocated as to cast them in a new posture in relationship to the black community, a posture that can ameliorate the alienation experienced by black youth and the emerging black middle class.

Gayraud Wilmore, Jr., in a 1986 article, "The Case for a New black Church Style," suggests that the Protestant denominations stop worrying about rapprochement with Roman Catholicism and begin dialogue on an unprecedented scale with the five major all-black denominations and the twenty-four smaller black churches that comprise more than ninety percent of all black Protestants in the Untied States. He emphasized that this should not merely be an exchange of fraternal greeting; rather, it should consist in joint planning for a total mission to the huge American cities having sizable black populations. Such a dialogue requires a respect for black theological thinking and style as the true reflection of the black religious soul.

America is in need of a new social revolution, one that focuses like a laser beam on the problem of racism. Such a revolution must be characterized by love, intelligence and sensitivity. It will, nevertheless, cause confusion, because the old ways of doing things, the old attitudes and prejudices will be challenged, and even attacked. Lerone Bennett Jr. has pointed out that it is difficult to tell time by revolutionary clocks because everything, including time, and space, changes in the revolution. In a speech delivered at Hunter College (1983) a few months before his assassination, Maurice Bishop, Grenada Prime Minister, defined a revolution as "a new situation, a break with the past," a situation in which the old order of things no longer occupies the same coordinates in the whirlpool of changing events.

The question, therefore, is not whether America needs or does not need a social revolution, but who shall lead the revolution, and what kind of revolution should it be. For the community most in need of a new American revolution, the black community, these questions pose a direct challenge both to courage and to ingenuity. The community is too small to make the revolution, but it has the genius to lead it. Actually, to the extent that the revolution has already begun, there is no question but that the community who first sang "We Shall Overcome" has already proven that it possesses revolutionary fibre. The sixty-four-dollar question is, what strategy will guarantee the needed participation of the white population, or at least a significant portion thereof.

CHAPTER ONE

IDEOLOGICAL AND LOCAL THEOLOGY

The theological pluralism with which we are so familiar today is a result of the mergence of multiple philosophies within and outside of the Catholic tradition. And the emergence of multiple philosophies is due to the growing determination on the part of peoples of various cultural and economic experiences to understand and articulate the faith from their own perspective. The single, unified framework of classical theology served well a society that shared the same ultimate presuppositions and philosophical assumptions, a type of society that did not encourage a philosophical and religious pluralism characterized by an appreciation of history, what Metz calls historical faith.

The classicist formally distinguishes between human nature's "natural" and "supernatural" end. For him, since the Church is the sacrament of the supernatural end, it declines to identify itself with any political entity. Thus classicism imposes artificial boundary lines between the natural and the supernatural which Metz finds both unacceptable and untheological. For Metz, to remove metaphysical limits is to introduce as an important element in theological understanding the secularity of the world in all its pluralism, a pluralism which Metz characterizes as "so vast that it can hardly be grasped as a whole."[13] While some theologians stress that an appreciation of secularity is a product of the modern age, Metz points out that it was also at the dawn of the modern age that the Church found suspect many manifestations of secularism within its own ranks. The Middle Ages theologians separated the *imperium* from the *sacerdotium*, defining the state as a secular creation of God, independent of ecclesiastical control or domination. Church and State, they declared, each makes its unique contribution to the bonum of the body-politic. They were more "modern" than the moderns. They understood what it means to be "secular."

---

13. Metz, p. 15.

The influence of secularism has so modified theological method that theologians of today cannot, without criticism, confine themselves to formulations of faith experience (*fides qua*) and of revelation (*fides quo*). They now investigate the relations existing between the many conceptual frameworks of today's pluralistic world, investigating and commenting on its many life forms. Metz explains that, in contrast to former times, faith and the world are no longer in a relationship seen and fixed by faith alone.[14]

The question of the compatibility of classicism and the new secularity of the world is an important one. There is a parallel in the question of the compatibility of Euclidean and non-Euclidean geometries. Our perception of space is Euclidean; we are certain that we can measure everything with a ruler. But space cannot be measured with a ruler, so these two systems are indeed incompatible. However, since both the Euclidean and the non-Euclidean perceptions arise from an individual's perception and understanding of space, they can be complimentary. The same is true in theology. Different experiences will form different theological categories, so the answer to incompatibility is the willingness to transcend individual perceptions. Then, what may in fact be incompatible can become complimentary.

To seek compatibility when complementariness will suffice is to posit an unnecessary confrontation. When the Jews complained to Jesus that his disciples were picking off stalks of grain on the Sabbath, Jesus cites the historical incident of David's friends who went into the temple and ate the holy bread because they were hungry. Of note is the fact that Jesus does not bother to justify the actions of His disciples. He simply cites the Old Testament incident without trying to artificially create an unnecessary incompatibility. For Jesus, it is the Sabbath that is relative, and human welfare that is absolute. "The Sabbath was made for man, not man for the Sabbath," (Mark 2:27). The law sometimes makes two compatible actions artificially incompatible. The action of His disciples was not "evil" simply because what they did was not compatible with the law. The secularity of the world can be made to be artificially incompatible, thereby establishing an unnecessary confrontation.

If classical theology stressed the importance of grounding the faith in doctrines that are not simply here today and gone tomorrow, today's theo-

14. Metz, pp. 42-43

logian stresses the fact that any particular theological formulation is finite. Artificial "confrontations" melt away with this understanding. Avery Dulles maintains that to describe dogmas as irreformable is an oversimplification of the reality of the theological discourse, that every dogmatic statement must be subject to reformulation if it is to reach ever deeper into its divinely revealed message.[15]

Dulles does not deny that both scholastic and more recent theological approaches have value. He would agree that, in the search for complementariness, one might consider scholasticism as a system, abstractive and logical, stressing what happens objectively (salvation or damnation), while current theology is existential and personal, stressing what happens subjectively (personal oppression, alienation, etc.). Complementariness is the key to harmony. Paul VI echoed Dulles during a General Audience in 1969 when he stated that we must make a clear distinction between the constitutional structures of the Church, which he encourages us to maintain, and those derived from historical tradition.[16]

John Courtney Murray, the American theologian, not only recognized that the secular and the sacred should be differentiated but went beyond complementariness to synthesis. He introduces a theological theme which we find later expanded by the liberation theologians. Commenting on the Vatican Council Declaration on Religious Freedom, he wrote in 1966 that its principle achievement was to bring the Church into the modern world, making it an active partner, together with other religious and secular movements. It succeeded, he points out, in clarifying the distinction between the sacred and the secular, and flushes out the new social synthesis this implies.[17] However, he did continue to emphasize the individual, and the Church's right and obligation relative to the individual. It is in the liberation theologian that we find a re-focusing of emphasis, this time on a society whose "community" is based on shared oppression. "Oppression" and "exploitation" become determinants in the formulation of theological methodology in South America. Gutierrez comments:

> A spirituality of liberation will center on a conversion to the neighbor, the oppressed persons, the exploited class, the despised race, the dominated country.[18]

---

15. Avery Dulles, *The Survival of Dogma* (New York: Doubleday, 1973), p. 165.
16. Paul VI, "Address to General Audience." May 7, 1969. Text in ***Osservatore Romano*** (English Edition), May 15, 1969, p. 12.
17. John Courtney Murray, "The Declaration on Religious Freedom," *America* (April 23), pp. 592-593.

In this statement the synthesis desired by John Courtney Murray is completed, as personal conversion translates into solidarity with our neighbor in the common pursuit of justice. The liberation implied in the Gospels, says Gutierrez, is not restricted to political liberation, but it does occur in historic happenings and in practical political action. In articulating his major themes, however, it is important to note that Gutierrez makes a great effort to assure us of his orthodoxy. He says that such critical reflection on the theological meaning of liberation demands, methodologically, that terms be defined and the critical function of theology sustained during one's exploration of the presence and activity of man in history. Classical theology called on us to reflect on the meaning and activity of God in history; a theology of liberation calls on us to reflect on the activity of man in history. Classicism stresses theological affirmations based on the Gospel message; liberation or ideological theology stresses the need to go beyond affirmation to action. To make an appeal for justice is necessary, but such an appeal is suspect in today's theological climate if it is not followed-up by a call to act to bring about justice.

Faith Includes Solidarity with the Oppressed

In Latin America, for the liberation theologian, faith must be understood on the basis of solidarity with the exploited classes. Theological truth emerges from a "real-life insertion" into the process of liberation. This personal experience is what differentiates the theology of liberation from other brands of theology. Quoting Bath, "Man is the measure of all things, since God became man," Gutierrez points out that today there is a greater sensitivity to the anthropological aspects of revelation, a trend away from being, per se, toward a focus on how man relates to his neighbor and the rest of the world. He calls for an awareness of the relationship between what is happening in the world and the formulations of our faith. In a genuine Christianity, the reading of the liberating Gospel raises consciousness, a consciousness that calls forth political action. The everyday life of man, as Christian, as he struggles with his social, religious and economic problems, is, for Gutierrez, a *locus theologicus*. He has been, he admits, influenced by Karl Marx, as he stresses that a study of the signs of the times isn't meant to be merely an intellectual exercise. Theological reflec-

---

18. Gustavo Gutierrez, *A Theology of Liberation* (Maryknoll, New York: Orbis, 1973), pp. 205-206.

tion must be, not an observation from above that remains abstract, philosophically theoretical, but a call to pastoral activity and commitment that relates directly and actively to man's life in the everyday world. This Marxist influence he acknowledges by quoting Sartre who contends that "Marxism as the formal framework" of modern philosophy cannot be superseded.

This stress on the activity of man in the world is a theological departure from Classicism, which tended to confine theological initiative to and examination of revelation and tradition. The scholastics drew very clear distinctions between the Church and the world. The "planes" are clearly differentiated, with church and state separated. The Kingdom of God provides the unity of God's plan; the Church and world contribute to its edification. This division left very little room for the priest and the dedicated layman to articulate a prophetic voice relative to social and economic oppression. This was Classical theology's major weakness.

For liberation theology, theological questions begin with the world and with history. Christians in the Third World have decided that they must come to terms with the political problems of oppression, and are developing a theology that is compatible with this activity. They have made commitments that have rendered obsolete the distinction of planes. The planes model is metaphysical, abstract and essentialist; the liberation model is historical and existential.

That theology can be both ideological and political is strongly affirmed by the Black theologian who articulates the hope and the suffering of the black community. It matters not to him that his theology is described as "separate" and "partial." In fact, he asserts his separateness from a classical theology which has not addressed itself to the black man's problems in an authentic and meaningful manner. He takes pride in his partiality, for it is the sign that he has taken up the burdens of his brothers and sisters in the manner of the Old Testament prophet. Segundo sees universality in black theology's "partiality." He sees the strength of the black theologians in their ability to adjust their ideological focus to the morality of oppression in the black ghetto. Their intent is to be contextual to black experience. That they can continue to do this without being exclusivist or elitist is their major challenge. Rosemary Reuther comments on this challenge advising that if the black theologian "wants its stress on 'oppression' to stand for universal anthropology" then the use of the racial metaphor must indicate that blackness refers to all oppressed people everywhere and the persons who seek their liberation. Maurice Bishop, of Grenada, understood this when he said, "Fidel Castro is the blackest man in the Caribbean."

In the context of theological method, this drawing of universal principles from a particular or local theology requires a methodology that, while not necessarily contradicting the classical method, differs from it in important aspects. Theologians from the black community have not been content to make their own the perspective of scholastic theology, without serious modifications. They insist that the experience of what it means to be black and live in the ghetto, and to have had a long history of slavery is essential both for the understanding of why a black theology is necessary, and the ability to contribute to it. Roger Haight sees this symmetry in relation to the South American situation, explaining that it is impossible to over-emphasize the importance of personal experience in the construction of theological concepts.

Local Theologies Always Existed

The motivating force behind the examination into this new theological method has been, primarily, the desire to develop a spirituality that has no difficulty in relating to and becoming a part of social action and political involvement, in areas where oppression and poverty result from unjust economic and social structures. Bernard Lonergan's *Method in Theology* and David Tracy's *Blessed Rage for Order* introduced new ideas and provoked much theological comment. While disagreeing with his style and language, Avery Dulles agrees that Tracy's theology reached out beyond a narrow circle of believers.[19] Nevertheless, he is afraid of Tracy's tendency to minimize the importance of the explicit Christian tradition.

While making the case for the new thrust in theology, we need to point out that in many ways, this "new" thrust is not new except in the sense that "local," historically, has referred to the authority of the majority, to the dominant culture's religious figures. Today it can refer to the religious influence of the marginalized, the numerical minority. Martin Luther King, Jr.,'s prophetic utterances dominated the genuine religious sentiment of the 1960's, even though his cause was that of the marginalized black poor.

---

19. Avery Dulles, "Method in Fundamental Theology," *Theological Studies.*

A theology which does not emphasize the evils of racism serves well a white society that is not determined to remove racism from its body-politic. Such a theology is found in the Afrikaner Church in South Africa. A theology, on the other hand, which focuses on the evils of racism is a theology which suits the hopes and aspirations of the black community, and so it is "local" relative to that community. The point I wish to make here is that those who consider it unorthodox to allow a proliferation of local theologies have themselves practiced and benefitted from self-serving local theologies. Local theology is practiced by oppressors and the oppressed alike. In the context of South Africa, Alan Boesak practices a theology of liberation and the Afrikaner Church practices a theology of racial superiority; both are local theologies.

Ideological and Balance Theologians

Theological nitpicking is a luxury we can ill-afford in an age that needs reconciliation, cooperation and mutual respect, but we do find it in the arguments for the pros and cons of the ideological and balance theologians. The so-called ideological theologians are identified as those who are said to take a frankly utilitarian attitude toward theology, an attitude that speaks of change, even radical change. There is also another frankly utilitarian attitude toward theology, one that is firmly committed to the economic and social status quo. Theologians of this school may not be called ideological but they are. The more moderate of them are called theologians of balance. They would, for example, accept the eschatological justification of suffering, while this is rejected by the ideological theologian. In this particular instance, the theologian of balance is confirming a long-held theological position, while the ideological theologian is taking a "new thrust." There is, therefore, a state of tension existing between the two schools, but this need not be seen as necessarily negative; tensions can be spiritually and intellectually productive and healthy. To deny the value of this tension is to create a dichotomy between the two groups which cannot be bridged. Langdon Gilkey believes that the deepest substantive question of current theology is the mediation of this false opposition, an opposition which he claims contradicts the biblical and theological interpretation of history and human destiny. A little humility on both sides could result in a much-needed integration. This integration, Gilkey says, will bring about a Christianity that inspires and shapes social reform and reconstruction, and make less likely our wilful participation in social oppression and disintegration. The balanced theologian's commitment is to the Gospel as attested by Scripture and Tradition, while the ideological theologian adds a specific commitment to human beings insofar as, being victims of structural oppression, they are somehow determined to free themselves from this oppression. Who can deny the value of each? The ideological or local theologian re-reads the Gospel from within the context of the liberating praxis that develops from this commitment. This is not to accuse the balanced theologian of somehow being short on compassion, but rather to question the appropriateness of emphasizing a study of the ultimate questions without particular and passionate reference to peoples who are oppressed or marginalized. The proliferation of local theologies is testimony to the fact that the "balanced" approach needs to make adjustments so as to address itself to the most pressing moral problems of our time.

Liberation Theology a Change of Focus

While classical theology seeks to respond to the challenges of non-believers who question the validity or even the existence of a religious world, liberation theology reflects on the person who is excluded from the existing economic, social, political and cultural order. This is a most significant shift of theological focus,[20] and was first noticed by black theologians in the work of Jurgen Moltmann. It is for this reason that Moltmann, whose existentialist theology helped prepare the soil in which liberation and black theology developed, has had such an influence on the black theologian's agenda. They saw in him an early hope that classical theology was prepared to open a dialogue with and address the needs of the black community. For Moltmann, the concrete utopia is realized only when man overcomes and destroys all racial identifications. The spirituality implicit in the liberation model stresses first that allegiance to God and to Christ is not and, more importantly, cannot be separated from a commitment of service to those who suffer from any form of oppression. This theological model also insists that there is an unavoidable relationship between salvation history and the liberation of the poor and the oppressed. Furthermore, this commitment to liberation is seen as a fore-

---

20. For a fuller treatment, see Chapter III of Juan Segundo's *Faith and Ideologies* (Maryknoll, New York: Orbis), 1984.

taste of that heavenly coexistence which most certainly is characterized by justice, equality and fraternity. When the black theologian demands the actualization of the liberation potential contained in the American Constitution, he sees himself as articulating a tradition of prophetic utterance that goes back to Old Testament times. That the black theologian might have delved even deeper into social analysis is testified to by James Cone:

> Another major limitation of Martin's and Malcolm's leadership was their failure to identify classism as a problem as harmful to the cause of freedom as racism and sexism...[21]

Secularism

Any discussion of the liberation implications of the Gospel message, especially if liberation includes the role of the Church in a revolutionary situation, leads to a discussion of the distinction between the secular and the sacred spheres of life. There is a deep-seated assumption in scholasticism that the secularism of the world is to be approached with great caution by the Christian. The pluralism of today's world and its highly articulated secularism is seen as a challenge to the compact, highly integrated Christian view developed in the Middle Ages, containing well-reasoned answers to all moral issues. The Christian community's response to the intrusion of secularity often takes the form of retreating behind the shield of eternal verities, insulating the faithful from its influence, and making it uncritically dependent on religious authority. Roger Haight observes that the world no longer functions on the basis of religious authority, and makes the point that liberation theology has integrated this fact into its discourse.[22]

The fear of secularism goes back to at least the period of the Enlightenment, for which "church" was synonymous with an oppressive *status quo*. That this "secular" period was creative in all areas of life, including religion, was not recognized by an officious and proud Church. The Christian community struggled with this period, which it found intimidating, and even now continues to struggle with the technological, sociological and geopolitical forces that have impinged themselves on the consciousness of orthodox Christianity; a Christianity that had neither the intellectual nor spiritual preparation to cope with the "winds of change." Canon

---

21. James Cone, *Martin and Malcolm*, (Orbis Books, New York, 1993), p. 280.
22. Haight, pp. 17-18.

Law remained unchanged from the fifth to the twentieth centuries, bringing about, in modern times, the call for an *aggiornamento* in church life. As painful as it was, this required an intensive examination of even the most sacred customs and beliefs. All of this had to be done in an atmosphere that began to question the isolation of the splendid superstructure of creed and liturgy, and began to attack what was called an artificial distinction between sacred and secular life. Langdon Gilkey comments:

> Our sin, our faith, and our obedience—are as constitutive of a social situation or a social history as they are of a personal, individual situation or a social history.[23]

The neo-scholastic nostalgia for the "religious age" remains the basis of conservative reaction to all attempts of the church to come to terms with the secular age. At the same time, on another plane, the forces of secularity naively suggest the option of a viable personal existence unrelated to church or religion. Secularism is an integral part of modern life. declaring that even immaterial values must be realized in time. "Secular" Christianity need not and in fact, for the most part is not, Christianity without God, or Christ, or the Church or worship, or the sacraments; rather, it is a challenge to the idea that religion has nothing much to do with the concerns of the world. In fact secular Christianity has stolen the fire of classical theology and declared that racism is America's greatest sin.

Christian Secularism Open to the Future

Christian secularism is open to the future, a future that is more humane, more conscious of human rights; the theology that has flowed from it is pregnant with the urgency of Gospel imperatives. Haight observes that in the modern sense of raised consciousness, modern theology, in contrast to the "finished" theology of scholasticism "is still what might be called a 'movement.'"[24] If black theology and liberation theology are still a "movement," then their articulation of the world and its secular meaning is too recent to be absorbed by a church that has formulated its concepts in the terminology of scholastic theology. And yet, the openness to the world on the part of Christians who are motivated to work for the issues of peace and justice gives hope that academic theology will not much

---

23. Langdon Gilkey, p. 114.
24. Haight, pp. 17-18.

longer lag behind. Certainly the intellectual and sociological content of classical theology has changed since Vatican II, and the growing influence of liberation theology and black theology will no doubt influence its justification even further.

The first major step in the development of flexibility and openness to the world in classical theology took place, not in academia but at Vatican II. The spirit of this openness is reflected in section 19 of the Pastoral Constitution, *Gaudium et Spes*:

> Hence believers can have more than a little to do with the birth of atheism. To the extent that they neglect their training in the faith...they must be said to conceal rather than reveal the authentic face of God and religion.[25]

In this passage it is clear that human beings share responsibility in common for the "God" which they project onto others. The very interesting assumption in the text is that Christians are actually responsible for the denial of God found on the sociological level. The actions of Christians, it more than implies, can conceal the face of God, and God Himself can be rejected because of the manner in which Christians appear to their brothers and sisters in society.

The Council Fathers also inspired today's theologians to come to an acute awareness of our world's structured injustice and institutionalized violence. Helda Camara, Archbishop of Recife, in Brazil, was unstinting in his praise:

> With the gospel message, the social encyclicals, Vatican II and Medellin, we have no need to appeal to any ideology to inspire us in our sacred commitment to foster human betterment.[26]

Vatican II's focus on salvation history was the product of trends in biblical study and biblical theology, but it made a dramatic contribution of its own. The Episcopal Conference of 1968 synthesized already-existing theological trends and expanded them. The latter constitute the post-Medellin theology that is now widespread. Theologians were encouraged to theologize in terms of distinctive historical and socio-cultural situa-

---

25. Vatican II, "Gaudium et Spes," No. 19; English Translation, Walter Abbot, ed., *The Documents of Vatican II* (New York: Guild America Association, 1966.
26. *CEIAM* (Monthly Bulletin of the Latin American Episcopal Conference), No. 127, July, 1976.

tions. The discussions that took place throughout an entire Church stimulated by the Conference is a clear indication that it had touched upon problems that go right to the heart of the Church's life. It was a flowering time for liberation theology. Major black theologians were emerging during this same post-Medellin period, independently of South American theological trends and influences. The black theologians, struggling to articulate the black experience, claimed for their people what St. Paul observed: "We are God's work of art, created in Jesus Christ to live the good life, as from the beginning he had meant us to live it." (Eph. 2:10). Both black and liberation theology have developed as critical reflections on Christian praxis in the light of the gospel message. And since that praxis in the oppressed communities of the world is increasingly one of liberation, the resultant theological reflection on it becomes theology of liberation. It is not a theology of politics or a theology of revolution, even though the reflection on liberation praxis must necessarily include political and even revolutionary considerations. And so it is that black theology, being a reflection on the social praxis as found in the black community, becomes ideological and local.

Enrique Dussel, the Argentinian theologian, finds that liberation themes go as far back as Exodus, and he discovers in Bartolome de Las Casas, the spirit of the liberation theologian. He quotes him addressing his fellow Europeans:

> Your own sinfulness is the cause of your present crisis. Now that you see you cannot exploit others endlessly and keep on eating off the fat of their land, you must begin thinking about living justly and humble.[27]

Local Theology Grass-Roots Based

The theology that has been emerging in the Third World (this would include oppressed minorities in the advanced nations) has developed with input from non-professionals. Liberation theology presupposes a role for the entire faith community in the development of its theology. We have seen this particularly in the *communidades de base* where the process of reflection in the application of Scripture takes place in the local church and, more precisely, in the local parish. The village of Solentiname, a remote archipelago on Lake Nicaragua, where the population of campesinos experienced theological dialogue with their priest in place of

27. Ibid, p. 207.

the customary sermon, is an example of theological refection that contributed to liberation theology. Copies of the gospel of the day were distributed to the people and it was discussed verse by verse. The purgoal was to understand the gospel message for that day in the context of one's daily existence and needs. Father Ernesto Cardenal, the founder of this particular Christian method relating to Church and Gospel, believed that the commentaries of the campesinos were of greater depth in analysis than that of some theologians, all the while having Gospel simplicity. Observe the theological analyses of the following Scripture quote given by one of the participants:

> *Text*: "And from now on all generations will call me happy, for Mighty God has done great things for me."
>
> *One of the ladies*: She says that all generations will call me happy...She feels happy because she is the Mother of God, the mother of Jesus the Liberator, and because...(she) understood her son and did not oppose his mission. She didn't oppose him, unlike other mothers of young people who are messiahs, liberators of their communities. That was her great merit, I say.[28]

One can see the themes of liberation theology weaving in and out of this response, and we can presume that none of the respondents had enough education to read the work of a liberation theologian. And so, when the liberation theologian states that the theology he writes is a reflection on the religious thoughts and the social conditions of the people where he lives, he is referring to this kind of grass-roots dialogue. Ernesto Cardenal, in December of 1977, two months after the village of Solentiname was destroyed by the Nicaraguan National Guard, wrote a letter to "the people of Nicaragua," explaining why he joined the Sandinistas. Contemplation, he said, had brought him and his colleagues to the revolution. Thomas Merton, his former novice master, had become his spiritual director, and the liberation movement in Latin America his obsession. Cardenal went to Solentiname in search of contemplation and he and his companions brought to that journey, initially, the classic notion of this discipline. After a time and after experiencing the oppression of the poor and politically marginalized, they brought not a new but a deeper meaning to the concept of contemplation. The desire for personal salvation became

28. Ibid, p. 28.

transformed into the desire for the liberation of the community in which they lived. And in the answers of the respondents of Solentiname we can observe that "community" is the concern of the people.

James Cone amplifies on his concept of theology when he observes that theology is the community's attempt to define in every generation its reason for being in the world.[29] Perhaps for too long we have taken for granted that the local community did not have a natural place in the process of developing theology. This is due to the elitist nature of the theological process, involving academics and advanced students only. Theology has been handed over to communities, fully formed *outside* the local situation. There is now a growing awareness that the context in which theology is done shapes reflection; the theological procedure follows the context. For example, in highly developed nations, where strongly individualist ideas prevail not only in the market place but in society as a whole, personal "inspiration" is emphasized in the community. In the less-developed world, where opportunities for individual development are much less and where interdependence of community members is great, theology based on community inspiration emerges. This is applicable to minority communities in developed countries. The black community in our nation is an example of where group action is a necessity, because of the limited availability of individual social or economic mobility. The "Harlems" of the United States are where the black theologians have sat and listened and protested with their people in the streets and in the political process, there being little or no distinction between spiritual" and "political" leaders. Perhaps one day all of theology will be the fruit of spiritual and intellectual dialogue between scholar and community.

Thanks to a new consciousness, we see theology being influenced by historical events, movements and cultures. The Jewish theology of today is influenced by the fact that the Jewish community has suffered so intensely over such a long period of time. And we see the emergence of a "woman's" theology due to the vocalization of long-suffered grievances. Women are reminding all of us that their sex has been discriminated against for a very long time, that the development of the church has been male-oriented. They are no longer waiting for an elitist theological com-

29. James Cone, *A Black Theology of Liberation* (Philadelphia: Lippencott Co., 1970), pp. 30-31.

munity to rescue them from their plight or take up their cause. Women writers and theologians of our time are a fine example of the making of "local" theology.

The term "contextual theology" has emerged to reflect upon and to attempt to specify the reshaping of a classical theology left untouched by these modern movements. Other terms such as enculturation, ethno-theology, theological identification, etc., are being used without a great deal of precision, as the meanings behind them shift and try to find a common context. One of the first terms used in this new perspective was, "indigenous theology," which emphasizes the fact that theology is done by and for a given geographical area, by local people rather than by outsiders. If the community becomes the center from which theology emanates, then theology can be seen as a teaching process resulting in the raising of consciousness among the people, in areas where action is needed in order to bring about liberation from a particular oppression. Such a theology is immersed in culture. Richard Niebuhr argues that "we do not know a nature apart from culture." Even Thomas Aquinas is a witness for the case that theology, if it is to be authentic, must come from a previous immersion in culture. In *De Malo*, 2,4 he argues that law is not the same in all places but differs according to the diverse circumstances of men and things. This is not to say that there are no problems in connection with "immersion in culture," that the apostolic tradition might not be diluted or minimized. But any dilution of tradition takes place not by praxis but by a mixture of local custom and outdated theology; the important corrective is to transmit church doctrine in such a manner that it relates to a cultural analogue.

Secularization

Johannes Metz not only had no fear of secularism, but even saw it as having a "Christological foundation," by which he meant that the entire history of Christianity has been imbued with the desire to draw the world into the saving reality of the Christ-event, with the determination to do away with the distinction between the sacral and the profane A secular orientation that brings the Gospel close to life addresses itself to this world, and is meant to be applicable to the needs of this world. The secularization of Metz is in contrast and opposition to a transcendental, existential and personalist theology that engenders an unhealthy concentration on the individual, private sphere of life. For the secularist, the Gospel is taken, not merely as a word addressed to the person, but as a promise given to man and society. We must conclude that the incarnation of God makes God an intimate part of history. He does not simply reign over the world; he also reigns within the world, a world of persons, not of things.

In America the secularization of the Church has taken place primarily in the black church which, because of the desperate social condition of its people, found it necessary to involve itself in community projects and political activity. This secular orientation in the black church manifested itself in the sermons of its ministers and their leadership in political affairs. The work and career of Adam Clayton Powell is an example of such secularistic tendencies. Tillich speaks of religion as being grasped by an ultimate concern, but this ultimate concern encompasses changing the earth. Harvey Cox defines "secularization":

> We have defined secularization as the liberation of man from religious and metaphysical tutelage, the turning of his attention away from other worlds and toward this one.[30]

Gutierrez judges this definition as an "early" one, good but incomplete. He prefers to understand secularization as the "result of a transformation of the self-understanding of man," or the condition in which man becomes an agent in his own history. If secularism is an index of growing self-understanding, then theology must reflect this new awareness by reflecting on man's acts and his experience in the world. This way of viewing theology helps in disabusing the academic community of the idea that only professional theologians determine the content and scope of theological reflection. It will remain true, of course, that the articulation of the local experience is going to be done by those in the community who, by virtue of academic expertise or natural talent, are best qualified to do this. But, even though this is necessary for intelligent articulation, the final "authorship" of the theological reflection remains in the community itself, because they must receive the finished product as authentically that which they have declared. The observations of the men we have quoted come after a long period of gestation. What causes a theologian to begin to think about theology as a tool of the liberation process? Bonino traces the development to a gradual awareness leading to charitable concerns, followed, in

---

30. Harvey Cox, *The Secular City* (New York: The McMillan Co., 1965), p. 17.

time, by an awareness of the need to build new political structures.

There are those who fear that such an approach to theology is dangerous because it substitutes the needs of a concrete present for that more universal, unchanging reality which the Church has always seen itself as representing. Such criticism fails to note that perhaps what was always considered to be the Absolute has been more a question of absolutes. If our gaze is fixed on the Absolute then theology can avoid the very narrowness which the critics fear. Liberation theologians "accuse" scholastic theology of having been so concerned with the absolutization of its own conclusions that the will of the Absolute has somehow become blurred. Preoccupation with absolutization leads to a kind of "establishment" guarding of its own prerogatives, to the detriment, at times, of the common good. Preoccupation with fixed parameters can cause the theologian to be unaware of changing realities. It is clear from the theological literature of today that it is the plight of the oppressed, not the plight of the nonbeliever that is the focus of attention. And, in America, the black man symbolizes the oppressed. Cone protests that the sin of American theology is its pitiful lack of passion. Even when it speaks of the poor it does so in language so languid that it implicitly discloses whose side it is on. Since passion generally is associated with self-interest, it may be too much for Cone to expect to find passion in the white theologian's analysis of the oppression in the black community. Nevertheless, Cone's remarks do reflect a reality in the theological literature outside of the black theological community.

The Value of Experience

Lack of particular experience, as seen from the perspective of the black theologian, limits the effectiveness of scholastic or classical theology. Experiencing oppression is the key to understanding the need for a theology that speaks to the morality of oppression, of finding ways to change structures that are so fixed they no longer produce a good life for the many. Black theology seeks to help liberate black men and women from the conditions in society that the white community has so absolutized. The Bishops at Medellin gave them hope when they declared that "We are at the beginning of a new historic epoch." The new method in theology, based as it is on personal experience with the lot of the oppressed, inevitably involves a transformation in the spirituality of the theologian himself;

this leads us to reflection on a spirituality which arises from a specific interest in the liberation of peoples. One man who has focused on this area of spirituality is the Chilean priest, Segundo Galilea. He seeks to develop a spirituality of liberation, and speaks of five intuitions that are prerequisites for such a spirituality. First, one's commitment to Christ cannot lay outside the service of commitment to the oppressed in society. Secondly, there is an intimate relationship between salvation history and the liberation of the poor, so that in working for the poor one is working for Christ and cooperating in His saving work. Thirdly, this work on behalf of the oppressed is the groundwork for the coming of the Kingdom of God, since it promotes justice, equality, fraternity and solidarity. The fourth intuition is the desire to see the oppressed liberated and to engage in the liberating praxis itself, a Christian charity incarnated in history. The fifth and final intuition is that of voluntary poverty, which produces a close personal identification with the oppressed. This focus on identification with the plight of the oppressed points out the fact that the eschatology of a liberating theology will be one that is grounded in the present, because no eschatological perspective would be sufficient which does not challenge the present order. If words so distort reality that they call for the oppressed to accept their suffering in the name of piety, religion or the salvation of their souls, then it is not Christian contemplation. James Cone puts the matter succinctly:

> What good are golden crowns, slippers and white robes or even eternal life, if it means that we have to turn our backs on the suffering our own children.[31]

Moltmann and the Theology of Hope

Since identification with the poor and the oppressed, and more specifically the black oppressed in America, is not an expected experience of the European theologian, it is not surprising that in black theology, the influence of Europe is limited. However, one European theologian who has impacted on black theology is Jurgen Moltmann. He insisted that only in a humane society where men no longer identify and differentiate on the basis of race can the future be attained. Utopia cannot be envisaged without a new identity of man that overcomes racial identification. There are cultural and theological differences between the Theology of Hope as pre-

31. Cone, *A Black Theology*, p. 241.

sented by Moltmann, and black theology as represented by Cone, but they both insist that God is at work transforming history and the present into the future. Moltmann's study of the ontological priority of the future as the mode of God's being, and black theology's realism grounded firmly in the daily experience of a sojourning people in a hostile land, combine to make a major contribution to the development of ideological and local theology. The eschatological emphasis of the Theology of Hope, however, is one of the points of departure between black theologians and their European counterparts. For Cone, the eschatological promise of heaven inadequately explains the historical pain of black suffering and cannot lead to accepting a God who inflicts suffering on the black community for some inscrutable purpose.

The Marxist Connection

The acceptance of the fact and the value of local theologies requires a conversion on the part of those who are accustomed to viewing theology monolithically. It requires a certain "deuniversalizing" to re-focus one's theological reflection on particular peoples and particular places. The Bishops at Puebla called for the church to experience conversion by adopting the preferential option for the poor. The conversion must contain within itself the willingness and ability to absorb into our total understanding of the theological needs of today's world the contributions made by non-Christians. Liberation theology is often accused of using Marxist analysis. This was inevitable, since Marxist sources helped many liberation theologians understand better the fact that religious forms of expression are intimately entwined with the totality of other social relationships. Marxist analysis has given clear articulation of the fact that some sectors of society have a special interest in the status quo, and that a new pattern of social relations must emerge before more than just superficial changes can take place in the pitiful condition of so many peoples around the globe. This is not to deny that there are some serious weaknesses in the Marxist analysis, for example, that economic conditions unilaterally determine religious consciousness. But even this exaggeration has prompted theologians to see the necessity of demythologizing popular religions. Haight observed that the historical and social analysis underlying Marxism is now being expressed in the context and language of liberation theology. The Marxist influence in today's political theology is obvious and is found (even if unconsciously) in the modern determination of religious communities to engage in social works of all kinds on behalf of the poor and the marginalized. In the new theological climate the credibility of religious communities depends on the contribution they are making to the alleviation of poverty and ignorance among the poor. Marx contributed the important concept that society and culture are constructed by human beings and are not a result of some inherent law of nature; hence, the form and structure of social life can be changed by human beings. If this were not so, then the poverty and oppression found in the human family would be accepted as unchangeable. Gutierrez acknowledges Marx's influence in theological thought in the understanding that theology must deal with the transformation of this world. It is necessary that liberation praxis continue; without an ongoing institutional and ecclesial praxis, along with continual social analysis, ideological abstractions could begin to stifle the most energetic communities. Those religious communities that have been influenced, in part, by Marxist analysis have managed to avoid the mistake of their political counterparts, that of viewing themselves as a vanguard party, an elitist group that presumes to speak for the rest of society. Among some who have worked with the poor and who are concerned about the economic as well as the "spiritual" welfare of marginalized peoples, there are some who strongly oppose the Marxist influence. Virgilio Elizondo, a member of the Ecumenical Association of Third World Theologians, and considered the leading theologian of the Mexican-American community, warns that the movement toward Marxist thought could result in trading one form of enslavement for another. He goes on to say, however, that the evil effects of liberal capitalism must be confronted. Bonino has no fear of Marxism; he contrasts the two realities (Christianity and Marxism) without antagonism:

> The Christian will, therefore, understand and fully join the Marxist protest against the capitalist demonic circle of work-commodity-salary. But out of the justification by faith alone, he will have to ask whether alienation does not have deeper roots than the distortions of the capitalist society.[32]

The very attraction to Marxist analysis on the part of most of the liberation theologians testifies to the validity of its being characterized as an

---

32. Jose Miguez Bonino, *Doing Theology in a Revolutionary Situation*, p. 111.

ideological theology, since they live with the people who struggle for liberation. Liberation theology is a theology with a cause, and in the area of black theology the cause is the liberation of the black man from his marginalized condition. Black theology does not hesitate to state that it intends to use religion in the service of liberation, for it is committed to the destruction of caste in America. Black theology is a sign of the times. It signals the most serious and effective engagement with white racism ever to take place in this nation. The particularity and value of black theology will endure until there comes about, in justice, a common religious experience in American society.

Black theology is an invitation to classical theologians to address themselves to the question of God's role in the contemporary American race issue. To the claim that ideological theology fails to deal with what is ultimate and transcendent, the black theologian answers that for finite man universal norms have limited applicability. At the same time, black theology is very cautious in its approach to the use of Marxist analysis because Marxism is, historically, so directly associated with antireligious and oppressive societies. So, while claiming itself to be ideological, it has been slow to absorb Marxist analysis with the same fervor as the liberation theologians of Latin America. In fact, black theologians accuse the liberation theologians of having adopted the cosmopolitan habits and customs of the Europe where they were trained, and minimizing the importance of popular culture and religion. Latins, they say, belong to the dominant cultural group in their respective countries, while the black theologians belong to a degraded American cultural group. The black theologian is more likely to stress the manner in which culture and religion can resist oppression, without the need for "foreign" ideologies. Black theology does not want to join the "revolution" and then discover that it has not addressed itself to the particular needs of the black community. Black theology, as local theology, hopes to enable the black community to see that the Gospel is commensurate with the achievement of black humanity. And in its affirmation of black humanity it provides authentic freedom of conscience to both black and white alike. Black theology affirms the humanity of white people in that it offers them the hope of conversion from their attachment to the sin of racism. It offers itself as a "local" theology to all of white America, a local theology which will offer an alternative to traditional American theology. Cone believes that white American theology, from the Puritans to the death-of-God, has succeeded in soothing the conscience of the black man's oppressor, and has even served as a divine sanction for slavery.

Black Theology Influencing the Dominant Culture

That a local theology which is not the theology of the dominant culture can strongly influence that dominant culture theologically appears to be a real possibility in American society. If we consider the civil rights movement as having sprung from black theology, then we can say that this has already begun to happen. But for this cross-cultural, cross-theological influence to spread throughout the American body-politic much more dialogue must still take place. At this stage in the development of black theology, dialogue is not easily attained. We know that there are some black theologians who do not think it necessary or desirable to discuss the nature of black theology with the white theological community. Other black theologians do not object as strongly in principle but have made the judgement that the present time is premature.

In spite of the problems that remain, there are signs that the sectarian exclusiveness of the black theologian is gradually diminishing. Perhaps the time is not far off when a more meaningful dialogue with the white theological community will be possible. This will require a corresponding willingness on the part of the white theological community to take seriously the academic work of the black theologian. As a doctoral theological student at Fordham University, my decision to develop a doctoral thesis on black theology was met with vacant stares by Fordham's theological faculty who, either directly or in effect asked: "Is there such a thing as black theology?" A theological faculty of a major university located in the same city that is home to the largest black community in America, and to major black theologians, yet so ignorant of black theology, is striking witness to the neglect attributed to the white theological community by black theologians.

Black theology has contributed to America's understanding of the plurality of experience, and the need to find appropriate categories which can incorporate each experience meaningfully. No longer is white history and white religious experience alone taken as normative. As a local theology, black theology has also contributed to the perception that theology is more effective when it is engaged with the most important moral issues in a society. There are limits to the usefulness of such categories as "oppressor" and "oppressed," but they have contributed to a better understanding of racist tendencies and structures in American society. The passionate and dynamic language developed in the black church has become standard American protest and civil rights language. Black theology is helping to shape the conscience of America.

Cone's invitation for white theology to "become black," i.e., to undergo a change of heart, and a change of theological method that will enable theologians of every color and nationality to engage in a common Christian project in favor of the oppressed and the marginalized, is promising. Black theology offers to those who wish to "do theology" in America, some relevant theology to do.

The Difficulties With Local Theology

Having discussed the many reasons why ideological or local theology is a natural outgrowth of reflections of Christians in a given faith community, it is appropriate to look at some of the difficulties it poses for Christian unity. The need for particular peoples to find their own identity before they can join in unity with others is one concern. This is particularly true of the black community in America after so many years of trying to conform to the dominant white culture, but it is also true of other poor, marginalized peoples everywhere. Nevertheless, while seeking self-identity and the resolution of local problems, the need for unity with the larger Christian community is never far from the minds of those who are doing theology in particular Christian communities.

Local theology goes beyond tradition in an attempt to be relevant to a given community, and yet tradition is of the very essence of a viable theology. One must try to remain faithful to the apostolic tradition while witnessing to the gospel in a given historical context. The problem is complicated when one has to deal with and study within the context of a demythologizing process that is on-going in theological literature and praxis. Demythologizing is particularly painful in the black community where the clinging to everything "black" is itself a protest against having to survive, culturally, for so many years, in an "alien" milieu.

Of all the "tradition" that has been handed down to the Christian community, which is the *real* tradition? And how can the real tradition be assimilable in various cultures? The liberation theologian has stressed how very important it is that any theological explanation be understood by the people to whom it is addressed. Pastoral practice has shown that particular peoples can accept the christian message while at the same time selectively filtering out parts of that message that reflect values alien to the local culture. Local theology, then, has built into it a natural confrontation with "tradition" as that word is commonly used in ecclesiology. The practical consequences of local theology's selectivity can be seen in Cone's assertion that a Copernican revolution has occurred within the black community, characterized by an attitude that societal values are no longer important. If "societal values are no longer important" in the black community, and if basic communities (*communidades de base*) engage in infrastructural support for guerilla warfare, then the attempt to sustain a unified tradition becomes extremely difficult. It was the hierarchical Church that encouraged the development of various forms of community organizations that would be resistant to both individualism and collectivism, the latter usually identified with Marxist socialism and the former with liberal capitalism. The rationale for establishing such community groups was the principle of "subsidiarity" which discourages assigning to higher levels functions that can be performed at lower levels. Avery Dulles describes such communities as "an alternative or complement to parish life." Dennis P. McCann cautions that what starts out as an attempt by the Church to encourage the development of local responses to unjust structures along the lines of traditional Christianity, soon becomes transformed (in the Latin American situation) into a political option for socialism. For McCann, the full meaning of liberation is possible only in Christ; no Christ, no liberation theology.

Political Theology

Political theology develops out of the hope that a new social *kairos* may be established that will offer many possibilities for liberation. Since black theology declares itself a political theology, a brief discussion of its nature is appropriate. The term "political theology" is relatively new but the concept is as old as the most ancient societies that uttered its philosophical and theological underpinnings. Political theology as proposed by Johannes Metz and Jurgen Moltmann purposely avoids new forms of theological support for the existing power structures precisely because it judges them unjust, not out of a hesitation to become involved. Metz defines political theology as "a critical correction of present-day theology" which places far too much emphasis on the private person rather than the society as a whole.[33] Metz's "critical corrective" refers to the need to remove faith from the private sphere and introduce it into the historical process. This relegation of religion to the private sphere applies not only to the classical theologians but to the transcendental, existentialist and personalist schools of theology as well. In fact, Metz goes so far as to say that these schools of theology might even have originated as a reaction to the situation created by the Enlightenment in France, which shattered the unity and coordination of religion and society, a unity which had been taken for granted. He insists that the story of Jesus in the Gospels is not biographical in the current sense of the word, but belongs to the genus of "public proclamation of kerygma," not to the intimate sphere of the individual. The task of the theologian is clear; he must engage in a political theology that has as its goal the deprivatizing of theology. Unless this deprivatizing takes place, Metz sees the danger that God and salvation will relate only to the existential problem of the person, thereby minimizing the eschatological kerygma to the point where it is reduced to a symbolic paraphrase of man's personal private decisions. In a broadening of the term "individual," Metz agrees with the emphasis on the individual in the New Testament and claims that the tendency to privatization causes us to bypass the individual in his real existence, for the existence of the individual today is entangled in a multitude of societal problems. This deprivatizing of theology is, for Metz, the negative function of political theology. Its positive functions is to "determine anew" the relation between religion and society, between eschatological faith and societal life. It is impossible to privatize the eschatological promise of liberty, peace and justice and any reflection on these major challenges of Christianity force us back, again and again, to the public forum. Every eschatological theology must be a political theology.

In all of this discussion there is introduced variants of meaning for long-cherished expressions and, admittedly, this can be confusing. The word "faith" itself does not escape modification. Political theology does not add something new to "faith"; it gives to faith the same parameters as man's historical activity. Hugo Assman maintains that the political dimension of faith is not something added to faith but faith itself in a particular historical context.[34] Assman portrays Jesus and the prophets as opposed

---

33. Johannes B. Metz, *Theology of the World* (Herder and Herder, 1971), p. 107.

to the legalism of orthodoxy and favoring an orthopraxies of truth concertized by political action in society, with the purpose of bringing about the future. For Metz, the transcendence of God consists in His calling us forward in terms of historical and political events, a constant uprooting and reshaping of the human society, a process of deeper and more meaningful humanization of history. Political ideology asserts that the Church herself lives under the eschatological proviso; she is not for herself but serves the historical affirmation of the liberation (salvation) of all men. She is the lamp of liberty in a world where self-interest necessarily militates against the freedom and liberation of our neighbor. Political theology sees the Church as a liberating force in regard to the social world and its historical process, and it sees its own task as the deprivatizing of theology. This is its negative task. Its positive task is to determine the relation between religion and society, between Church and societal "publicness." In America, black theology has been busy doing just that.

Summary

In Chapter One, I have attempted to describe the process of making theology that is both ideological and local. I felt it was important also to establish the legitimacy of such a process since black theology is both ideological and local. I have also tried to show the manner in which ideological theology contrasts with and is critical of classical theology. I pointed out that ideological theology is not something new but has its roots in the earliest societies of mankind, and today plays the role of "critical corrective" as it introduces faith to the historical process. In Chapter II, I shall describe the roots of black theology as the milieu in which African slaves worshiped and expressed their ideas of God and the world and the relationship between the two. I shall show that black religion (and the theology which articulated that religion) was a religion of protest, a protest against the dehumanizing effects of slavery. It was under these circumstances that the black man felt the need to create a theology of his own. If white Christianity could so dehumanize both the black man and the white man, then a new expression of what it means to be a Christian was needed.

---

34. Hugo Assman, *Theology for a Nomad Church* (Maryknoll, New York: Orbis, 1976), pp. 34-35.

CHAPTER TWO

SLAVES AND THE DREAM OF ABOLITION

The death of Dr. Martin Luther King and the publication of *The Black Messiah* by Rev. Albert B. Cleage, Jr. represented the end of religion-as-protest, and the beginning of the era of religion-as-reconciliation. Dr. King held firmly to his vision that black and white Americans could and must live in harmony. He hoped to demonstrate that nonviolence could successfully attack structural racism and provide a valid alternative to violence. Albert Cleage's work represented the flowering of a black theology as theological rationale for the goals and strategies of black liberation. Cleage worked to assimilate the radical, anticlerical elements of the black power movement. It is true that his work restructured a black church which remained strongly influenced and true to elements of white church worship, but it also blunted and curbed the radical, violent-prone movements which were the inheritors of a protest movement with deep roots in the ante-bellum South. The foundation of black religion in America, conceived against a background of slavery and segregation, was an angry religion. Martin Luther King, Jr. desired to transform that anger into meaningful non-violent strategies that would result in changing racist America into a land of brotherly love. He was not, as some of his angry co-workers believed, a man who was eternally naive. He analyzed, as did Gandhi, the social and political situation of his native land, and decided that this strategy was the strategy for his time. He was the prophet who both spoke of injustice in fiery terms and promised to those who patiently waited and prayed with him, deliverance. He refused to accept the proposition that man is "mere flotsam and jetsam" in the harsh realities of life around him, that a universal brotherhood could not become a reality.[35] King's optimism was not shared by the African slaves on the plantations.

---

35. From Martin Luther King, Jr.'s "Nobel Prize Acceptance Speech" as quoted by Flip Schulke, ed., in *Martin Luther King Jr., A Documentary —Montgomery to Memphis* (New York: W.W., Norton, 1976), p. 219.

They saw no possibility that their Masters would one day be so kind as to consider them as brothers. The religion of the slave was not the religion of the slave master, except externally when the slave master built a chapel on the property and required the slaves to attend services with him. The slaves sang and dreamt of an elusive freedom, while a few among them seriously plotted uprisings, even in the face of the most overwhelming odds and almost certain capture and death. Black religion became a religion of protest and future hope, as plotting and prayer went hand in hand. Nowhere is this more in evidence than in the music of the slave which, comparing the slave condition with the Hebrew bondage in Egypt, calls on the authorities to "Let my people go."

The Black Church Is Institutionalized

These expressions of protest and future hope became institutionalized gradually, as the small group meetings of the slaves began to form the first black church in America. The formation of this black church had its own definite historical foundations. These slaves came from a social environment characterized by a polygamous clan life under the authority of the chief and the priest. His worship consisted in incantation and sacrifice. His forcible insertion into the slave ship and the West Indian sugar cane fields threw his life into utter and complete turmoil, as the plantation replaced the clan. The forced labor which never seemed to end became the rule of his pitiful existence. The old ties of blood relationship and kinship disappeared, as family members were scattered all over the West Indies and the Southern States. The African family was what anthropologists call the "extended family," with two or more generations of adults sharing a compound consisting of a group of dwellings enclosing a courtyard. This very secure existence was replaced, in slavery, by a new kind of uncontrolled polygamy, as men and women torn from their roots and bowed by toil and repression, sought out each other for love and comfort amidst their misery and pain.[36] The chief had disappeared but the "medicine man," or priest, was transported to these shores with the others and remained in their midst. He became the healer of the sick, the comforter of the sorrowing and the one who most eloquently expressed the bitter resentment and longings of an oppressed people. The African "priest"

36. A good description of African family life is given in August Meir's *From Plantation to Ghetto* (New York: Hill and Wang, 1976), especially in Chapter One.

became the Negro preacher and it was under his tutelage that the first Afro-American institution, the black church, began to grow. Under the veneer of Christianity, African rites and practices flourished in this early church. The African background was important in the choice of Christian denomination. The slave gave his adherence to the Christian sect whose ritual most resembled the kind of worship with which he was familiar. And all the while that the small meetings, usually in secret, developed into congregations of worshipers, the theme of freedom dominated the liturgy. W. E., Du Bois observes:

> For fifty years Negro religion thus transformed itself and identified itself with the dream of abolition, until that which was a radical fad in the white North and an anarchistic plot in the white South had become a religion to the black world.[37]

The Newly-Freed Slave

The desperate circumstances of the slave and the newly-freed slave engendered emotions of anger and rage that erupted in the "Harlems" of the North, where the former slave went to seek freedom. Du Bois sketches the building-up of anger and radical complaint in the young black who, socially dwarfed since childhood, seeks a new-found freedom that both implodes on his psyche and explodes on the society around him. The anger of the slaves continues to linger in the black community. The black church was born in slavery and its existence symbolizes a people who were completely stripped of their African heritage as they were enslaved by the "Christian" white man. The slave was a *no-thing* in the eyes of the Christian master, who did everything possible to instill this sense of nothingness in the mentality of the slave. It should not surprise us that if the white man who whips and rapes a black woman worships God, that "God" may be, for the black man, a source of alienation. In 1973 the Caribbean Conference of Churches published a collection of essays entitled *Troubling of the Waters*. Idris Hamid, who edited these essays, wrote that God is a foreigner to the Caribbean peoples, a guest the white man brings home. In the same selection of essays, the Reverend Ashley Smith, commenting on statements made by Walter Rodney, the important West Indian intellectual who was silenced by an assassin's bomb, had this to say:

---

37. W. E. Burghardt Du Bois, *The Souls of Black Folk* (New York: New American Library, 1969), p. 221.

> The intensity of black self-hatred, body-shame and fatalism is due mainly to the use of religion as an instrument for the inculcation of the 'white bias.'[38]

The "God" which Idris Hamid experiences as an outsider is the white man's God who must be, if one observes the action of the white man, a Non-God. But the real God of the Bible was very much believed in and prayed to and appealed to every day by the black slave and the black freeman.

Basic Themes

The divine liberation of the oppressed from slavery is the central theological concept in the black Spirituals. These songs show that the black slaves did not believe that God created Africans to be the slaves of Americans. Accordingly, they sang of a God who was invoked in history, their history, making right what whites had made wrong. Just as God delivered the children of Israel from their Egyptian bondage, drowning Pharaoh and his army in the Red Sea, He will also deliver black people from American slavery. It is this certainty that informs the thought of the black Spiritual. The basic theme is that slavery is in opposition to the will of God. To be enslaved is to be declared nobody, and that form of existence contradicts God's creation of people to be His children. Black people affirmed their "somebody-ness," refusing to reconcile their servitude with divine revelation or the norms of a humane, rational society. They rejected white distortions of the gospel which countenanced the obedience of slaves to their masters. They contended that God willed their freedom, not their slavery. Black people sang about Joshua and the battle of Jericho, Moses leading the Israelites from bondage, Daniel in the lions' den, and the Hebrew children in the fiery furnace. Here the emphasis was on God's liberation of the weak from the oppression of the strong, the lowly and down-trodden from the proud and mighty. And the slaves reasoned that if God could lock the lion's jaw for Daniel and could cool the fire for the Hebrew children, then He could deliver Black people from slavery.

Slave religion was also dominated by the theme of freedom-in-bondage. Sometimes the secret meetings of black congregations were the occasions for planning overt resistance. At other times the reality of freedom

---

38. Ashley Smith, in *Troubling of the Waters* (San Fernando, Trinidad: Rahman Printery, 1973), p. 44.

was affirmed in more subtle ways. The theme of liberation was so strong that slaveholders did not allow black slaves to worship openly unless authorized whites were present to proctor the meeting. The spirituals speak about the rupture of black lives; they tell us about a people in bondage, and what they did to hold themselves together and fight back. We are told that the people of Israel could not sing the Lord's song in a strange land, but not so for blacks; their being *depended* upon song. Through song they built new structures for existence in an alien land, enabling them to retain a measure of African identity while living in the midst of American slavery, providing both the substance and the rhythm to cope with human servitude. The New World African's accommodation of the slavemaster's religion and the retention of Africanisms produced the Jesus-faith of the antebellum slave which remains identifiable today. That Jesus-faith, which was preserved for posterity in the Spirituals, served to insulate the ante-bellum slave from the real temptation of collective suicide.

The Negro Spiritual

The Negro Spirituals fashioned in the slave warrens of the South authenticated the Black religious experience. Although when obliged to worship with their white slavemasters, the slaves participated in and learned verbatim the Euro-American hymns and their style of worship, they knew that the retention of their own music and traditions was vital to the preservation of their African identity. The Spirituals have a rich musical, historical and theological heritage. They are rich musically because they move the body to exciting and pulsating rhythms. They are rich historically because they chronicle the travels and travails of black Africans in slavery. They are rich theologically because they narrate the journey of the slaves with God and the saving presence and grace of Jesus in their midst. The Old Testament is much more extensively represented in the spirituals, for its stories of the Hebrews in bondage immediately resonated in the life of the slave. Africa also dominates the Spirituals. There was such a large number of West African tribes involved that a common linguistic tradition did not survive intact; the commonality of being African and being enslaved, however, was sufficient to forge a new and dynamic religion. This Afro-based religious development became the authentic American black culture and is found only in the black church. When a black choir sings about freedom, they do so from the knowledge that, conservatively, about fourteen million Africans (about three million in Brazil alone) were imported into the Americas during the Atlantic slave trade. For every African who reached these shores alive it is estimated that four died in the machinery of slavery during the dreadful Middle-Passage. This means that as many as sixty million Africans may have died during this time, a genocide on a scale unmatched in all of history.[39]

From about 1670, in the plantation colonies laws were enacted that made it illegal to permit a black slave to read or write. By 1700 America had developed also the legal apparatus that was to make black slavery the most oppressive form of slavery the world has ever known. The damage done by slavery went far beyond the physical. Recent cliometric studies of the American slavery system, while providing valuable insights into the material aspects of slavery, have been limited as to the kind of information they could provide on the psychological damage. Alfred Duckett comments:

> And, while the cliometricians have been able to construct a reasonably reliable index of the material level at which blacks lived under slavery, it has been impossible, thus far, to devise a meaningful index of the effect of slavery on the personality or psychology of blacks.[40]

The historical and mystical African heritage was all but destroyed by slavery, but there did remain one instrumentality that slavery could not destroy, music, which became the medium by which the yearning for freedom was articulated. That which the slaves dared not utter in public word, they disguised in song, thereby holding together their humanity and leaving the door slightly ajar for hope. This was the music of the ante-bellum slave. The Spirituals choreographed the dynamics of the slave community. They mirrored the social context in which they grew and developed and, because of the strict rules governing communication among slaves, music became the chief means both of protest and the handing down of the African heritage. Hildred Roach explains:

> The method of oral tradition was greatly responsible for the maintenance of the samples of African heritage which miraculously survived the centuries.[41]

---

39. For a modern work on Slavery, see *Negro Slavery in Latin America*, by Roland Mellafe (Berkeley: U. of Calif. Press, 1975), p. 70.
40. Alfred Duckett, "An Interview with Thomas A. Dorsey," *Black World*, vol. 23, no. 9 (July, 1974), p. 13

It is interesting to note that much of what America claims as its own music has come from the black man and is, therefore, linked to Africa. The sighs and moans of the slave became the stuff of which white America's music was made. John Rublowsky observes that much of our music came to America in the holds of the slave ships, a music with unique qualities of sound, melody and rhythm. The heavy concentration of biblical themes in black music is the earliest expression of the theology which was in the process of developing among the slaves. Nearly all the Spirituals are taken from biblical themes, but especially those themes in which by supernatural means God delivers his faithful from some form of bondage or danger. Because of the desire to avoid unnecessary confrontations with the slave master, references to the slave system are not made directly, but the veiled hope for the Promised Land is always present. Singing the Spiritual was a nonviolent way of protesting slavery. In our time this same nonviolent method has been used in the civil rights movement, symbolized by the song, "We Shall Overcome."

For those of us who have not personally experienced oppression, the Spirituals could be just songs sung in a church and having a biblical theme. But in these songs the black man sings of his history of slavery and oppression in the United States. Israel becomes the black race and Egypt is America. It's ironic to recall that those white Christians who practiced slavery found blacks, especially the uneducated, unworthy of entering a white church, much less being baptized. The ever-present contrast between white religion and black suffering convinced the black man that he had to remain faithful to his own religious roots, because it became obvious that the adoption of "white" theology or "white" religion by the black race would be nothing less than self-genocide. The question, for the slave and his descendants, is not whether black theology is valid or not; the question is whether the black race could survive without it. Looked at from another perspective, one can only wonder at the fact that the black man bothered to retain any interest at all in Christianity, that he was not scandalized to the extent of abandoning Christianity altogether. The only explanation can be that the black man has the ability to look beyond particular persons and events to the heart of the Gospel message.

---

41. Hildred Roach, *Black American Music: Past and Present* (Boston: Crescendo Publishing Co., 1973

Acceptance of Christianity Not a Sign of Submission

The ability of the black person to transcend the suffering inflicted on him by white Christians was due largely to the fact that the slaves and black freemen accepted the Gospel message. He/she did so, not because the slave master insisted, but because the Gospel and the personality of Jesus was enthusiastically received. All the evidence, from slave revolts to the content of the sermons of the black preacher, to the underground railroad, to the "invisible institution" indicates that conversion took place without submission rather than because of it. The slaves saw the inconsistency between the white man's claim that God was terribly interested in their eternal salvation and the powerlessness and wretchedness of their condition of slavery, imposed and directed by white Christians. The slaves were not interested in being "converted." This marketplace missionary effort was offensive to their sense of what religion was all about. In Africa, religion was not the manifestation of the desire to convert, but an atmosphere of joyous affirmation in the presence of the Almighty. Neither did the African make personal sin a part of his religion, and this puzzled the white missionaries to the slaves. To the European missionary who came from a long tradition of religious individualism and a strong emphasis on guarding against personal sin, the communal attitudes of the slaves toward everything, including religion, was difficult to comprehend. It was not part of the African tradition, for example, to "confess" during religious rituals. This kind of ritualistic celebration, with signs of repentance for sin noticeably absent, was a source of annoyance to the white missionary. It was in the thralls of slavery that notes of pessimism crept into the slaves' religious celebrations, and this note was communal also, not personal. Wilmore explains that the religion the slaves practiced developed out of the growing consciousness of their blackness and the realization that their color had something to do with their condition of slavery.[42] The extent to which the white owners and missionaries misunderstood what was going on in the hearts and souls of the slaves is an index of how little they knew about them. One such missionary reported:

> I was preaching to a large congregation on the Epistle to Philemon; and when I insisted on fidelity and obedience as Christian virtues in servants ....one-half of my audience deliberately rose up and walked off with them-

42. Gayraud Wilmore, *Black Religion and Black Radicalism* (Maryknoll, New York: Orbis Books, 1983), p. 13.

> themselves; and those who remained...solemnly declared that there was no such Epistle in the Bible; others...that they did not care if they never heard me preach again.[43]

The religion of the slave was a religion of protest, of revolution, of a determination that however long it took, they would be free. The black interpretation of Christianity is directly connected with the concept of freedom. Christianity liberates. Jesus saves. He saves the soul and He saves the body from bondage. Jesus is the liberator, the friend of the oppressed. The slaves were poor, oppressed and uneducated, but they sneered at white hypocrisy. They listened to white preachers and masters extol the virtues of meekness, humility and obedience, but they never bought it for one minute. Survival demanded silence, but that silence was sullen and can still be observed in the black ghettos of America. The black man worships in the black church, suspicious of every virtuous sermon and declaration coming from a white church which did not in ante-bellum times, and does not today preach with indignation on the evils of racism in America. Laurence Thomas, a feature writer for New York's most influential West Indian newspaper, wrote:

> The other-worldly approach was fine for white folks who were enjoying a taste of heaven right here on earth. It did precious little for Black folks whose lives were constantly being scorched by the hellish flames of racism....Racism is an evil, the utter demise of which every Black person has a stake in...[44]

The protest of the slave is echoed in the land today as black religion continues to protest against black suffering. This is not to say that every religious moment in the black church is pregnant with anger and protest, for there is a plenitude of joy and love and fervor. Although in the days of slavery religion was used as a coding system to keep their own covert and subvert meaning systems alive, it also served the purpose of worship and petition. In fact, this is what makes black religion so unique, namely that it combined the two then (in the days of slavery) and now. The slaves were deprived of their native land and their memories; they were demeaned and degraded. But, in the midst of all this suffering and degradation, imposed and self-imposed, there developed a religion of song and dance that has no

---

43. Wilmore, p. 9. Here he quotes the report of a C.C. Jones *(Tenth Report of the Association for the Religious Instruction of the Negroes in Liberty County, Georgia).*
44. Laurence Thomas, "When Slaves Get Together," *Carib News*, 21, Jan., 1986, p. 18.

equal among men anywhere. Without figurative straw, the slaves made the bricks of religious institutions. "Black church" and "Black religion" are metaphors for the shared historical and spiritual experience of black Christians in America, especially from 1787 to 1865. It is true that there were many white ministers who wished to proselytize among the slaves, and certainly a lot of this was done, but the black church was developed for blacks by blacks. Robert Anderson, an ex-slave, gives us the setting:

> Our preachers were usually plantation folks just like the rest of us. Some man who had a little education and had been taught something about the Bible would be our preacher...We had our meetings of this kind, we held them in our own way and were not interfered with by the white folks.[45]

The Persistence of Africanisms

Remaining "African" has always been important to the black man in America, from the time of slavery to the present. It was this determination that prevented the slave from disappearing from his own vision. This passionate focus on "Mother Africa" was reflected in the black church, and so a brief examination into that phenomenon is important. The debate over the extent and persistence of Africanisms in the religious life of the American negro has been a lively one. Melville Herskovits and E. Franklin Frazier, represent the two differing schools of thought on the subject. Herskovits, a white man, claims that black religion and customs in America are heavily influenced by "Africanisms," while E. Franklin Frazier, a black man, claims the opposite. Most scholars side with Melville Herskovits, including the best of the black scholars. Linguists, folklorists, anthropologists, and specialists in the history of music and dance have explored and given academic opinions on the subject. When all is said and done, there is a consensus that African cultural forms were not as overwhelmed by the slavery experience as was formerly believed. George Rawick observes:

> The black slaves in North America utilized West African concepts in a new and totally different context. In so doing, they transformed those West African forms into something which was neither African nor European-American, but a syncretic blend of the two...[46]

---

45. Robert Anderson, *From Slavery to Affluence: Memoirs of Robert Anderson, Ex-Slave* (Heminford, Nebr., 1927), pp. 22-23.

Rawick points out one cannot separate "African" traits or "European-American" traits from the resulting product, without violating the cultural combination. In any case, whatever one may conclude about the vestiges of Africanisms among slave populations, the fact is that it was the Christian religion that provided, in the United States, the new basis of social cohesion. This Christian religion, as many a missionary and parish priest was to find out, was imbued with the spirit of Africa and, in many cases, this fact was itself a form of protest against having to learn a new religion in a strange land, even if that new religion appealed to them and offered the promise of liberation.

Africa remains a deep presence in the black community. The historical circumstances in which religious traditions from Africa have been transmitted to New World societies varied from place to place and from region to region. African gods and rituals have been able to survive to a far greater extent in the Caribbean and in South America, especially Brazil, than in the United States. Most anthropologists explain this by noting that in the Caribbean the slaves came under Catholic influence, while in the United States the atmosphere was Protestant. In Haiti, Cuba, Brazil and some of the Caribbean islands, Catholic devotion to the Blessed Virgin and the saints offered a rich context for syncretism with the gods of Africa. Alfred Metreau reports that the clergy in Haiti complained about their parishioners, that they were mingling the sacred and the profane. This mixture of "the Holy Utensils of our religion with profane and idolatrous objects"[47] infuriated the priests. It never occurred to them that this was most natural. The use of sacraments, such as statues, pictures, candles, incense, holy water and rosaries was more akin to the spirit of African piety than the austerity of Puritan America, where such objects were said to be idolatrous. Holy days and processions were all easily adapted by Africans who had observed the sacred days, festivals and taboos of the gods. The more Catholic the country the slave was taken to, the greater the remnants of African tradition and lore. In Cuban religious processions, for example, the Negro cabildos played an essential role in spite of, or perhaps, more accurately, because of the thinly-disguised African fetishes dressed up as local saints and the virgin.

---

46. George Rawick in *From Sundown: The Making of the Black Community,* Vol. 1 of The *American Slave: A Composite Autobiography,* Series One (7 vols.: Westport, Conn.: Greenwood Publishing Company, 1972), pp. 37-38.
47. Alfred Metraux, *Voodoo in Haiti* (New York: Schoken Books, 1972), p. 101.

Africanisms As Syncretism

There has always been great resentment on the part of people of African origin that the white community constantly referred to African religious concepts in terms of "paganism" and "fetishism." The continued use of "Africanisms" amongst congregations of black religious organizations and churches is, in part, due to the same feelings of protest and outrage at white control and interference in their affairs experienced by their slave ancestors. They still point out that early European travellers to Africa frequently identified African gods with demons or devils and accused Africans of devil worship. Black Americans are outraged at such continued distortions. Even the representation of the gods as fetishes is a mistake, for a fetish is simply a charm or amulet, like a medal of St. Ann. Common to all African societies was a High God, or Supreme Creator of the world and everything in it. The lesser gods were not "god" in the divine sense but more like our saints, to whom we pay homage without fear either of offending our Almighty God or of divinizing the individual saint whom we honor. The Ashanti called these lesser gods *abosom*. The Ibo worship them as *alose*, and the Yoruba call them *orisha*. I mention this because, since the American slaves came primarily from these three tribes, these various "gods" did play a role in the early development of slave religion.

As in any other cultural assimilation process, the assimilation of the African to the American and Caribbean environment took time. At first, the African gods had to be identified in a Christian manner, since the slaves were obliged to be baptized and to follow, at least externally, the Christian liturgy of the slave master. The missionary in the Caribbean, if he spends any time at all studying the religious practices of West Indians, will soon become very familiar with the names, Eshu, Ogun, Yemanja and Shango. These gods were given Christian names in the slave situation. Shango became identified with St. Anthony. The Catholic notions about the role of Christ, Mary, guardian angels, and a patron saint as intercessors with the Father in heaven, for men on earth, proved quite compatible with African ideas about the intervention of lesser gods in the day-to-day affairs of human life. The rainbow god of Quidah, symbolized as a serpent, is sometimes identified with Moses, because of the miracle of the brazen serpent, and sometimes identified with St. Patrick, pictured driving the snakes from Ireland. Ogun becomes St. John the Baptist, Oshun becomes St. Lazarus, the beggar covered with sores. Albert J. Roboteau describes the Shango cult ceremony in the Caribbean:

> Correlation of saints with gods is only one aspect of syncretism between Catholic and African forms. Candles, crucifixes, and chromolithographs are blended with rituals associated with African gods. In Trinidadian shango the annual ceremony of the shangoists...begins with prayer meetings in which an incense burner, lighted candles, Catholic prayers...In *candomble* the drums themselves are baptized in the presence of godparents according to Catholic ritual...cult members are simultaneously worshipers of the African gods and communicants of the Catholic Church.[48]

The contrast between Catholic Caribbean lands and Protestant America as an explanation of African retentions should not be pressed too far. In Jamaica, the Cumina and revival groups are an example of synthesis of Protestant and African religions, and in Louisiana Afro-Americans were able to preserve only a vestige of the African rituals, in spite of the fact that Catholicism was the dominant religious presence during the periods of French and Spanish rule. But, the general rule of thumb does hold. Of the many important factors inhibiting the survival of African culture among slaves in the United States is the fact that, in sharp contrast to the Caribbean, the slave population increased at a rapid rate, increasing the native-born slave whose memory of the African past grew fainter with each passing generation. In the Caribbean, the slave population steadily declined and the retention of vestiges of African culture remained strong among a tight-knit group. Another interesting factor is that the number of slaves brought to the United States was very small and they were scattered over a very wide geographical area. The opportunity to hold a culture together under these circumstances was minimal. The Greater Antilles has an area roughly one third that of Texas, but it imported nearly six times the total number of slaves landed in the United States. The island of Saint Dominique alone imported 864,000, which is more than twice the United States total. Cuba took in more slaves after 1808 than the United States total. The relationship between the small amount of slaves imported into a Protestant America and the large amount imported into a largely Catholic Caribbean and parts of South America is important because it lays the basis for understanding why black religion developed first in the United States. Here, the African, having lost a good part of his customs early-on, embraced Christianity with much less syncretism of cultures than he did elsewhere in the Americas. With a relatively small amount of slaves scattered over a very large geographical area, in an atmosphere of Puritan reli-

48. Albert Raboteau, *Slave Religion: The "Invisible Institution" In the Antebellum South* (New York: Oxford University Press, 1978), p. 24.

giosity, it was not possible to maintain the rites of worship, the priesthood, or the national identities which supported African religion and cult. However, the African heritage of singing, dancing and spirit possession did continue to influence Afro-American Spirituals and beliefs. Henry Mitchell, in speaking of African survivals among American slaves, had this to say:

> To kill a culture you have to kill the bearers of that culture... pressure to destroy them often succeeds only in driving them underground...a people's culture is transmitted from "unconscious" to "unconscious." To stamp out a culture and its world view would require total genocide.[49]

Ms. Norma Blaize, former Grenada diplomat and member of the Big Drum Nation Dance Company, of Carriacou, Grenada, in the West Indies, speaks about the importance of African retentions:

> The Big Drum Dance and the feasts at which the dances take place, remind us of our African origins. It is not merely a question of dancing because the Big Drum is associated with the African spirits. For Example, I'm an Ibo and when the Ibo dances are done, I participate...we set aside a room in the Catholic Church and...put food for the spirits of our ancestors.[50]

Father James, the former Episcopalian priest on Carriacou, issued a newsletter, *Shells*, which he sent to friends abroad. In the July 1985 issue he had this to say:

> From the first days since the white man came to these shores...either the Roman or the Anglican Church... was used by the Crown to domesticate the slaves. It served the Empire by making the slaves Christians... The politics and the church is colonial, not West Indian or West African. It certainly is not black...

In the "Harlems" of America, as in the West Indies, the psychological and spiritual problems caused by colonialism and slavery remain. It is for this reason that prophets in the black community have striven mightily to rid themselves and their flocks of the "slave mentality," and it is for this reason that the black church remains a church of protest and subdued anger. Joseph Washington comments:

---

49. Henry Mitchell, *Black Belief: Folk Beliefs of Blacks in America and West Africa* (New York: Harper and Row, 1975), p. 1.
50. A conversation with Ms. Norma Blaize, of Carriacou, Grenada, in Brooklyn, January 27, 1986.

> ...the uniqueness of black religion is the racial bond which seeks to risk its life for the elusive but ultimate goal of freedom and equality by means of protest and action.[51]

Black theology has had to deal with the fact that from the very beginning of the Atlantic slave trade, conversion of the slaves to Christianity was viewed by the European nations as a justification for the psychological as well as physical enslavement of Africans. Whatever guilt feelings these Christian slave masters may have entertained were assuaged by pointing out the grace value of the faith made available to these "pagans."

Black Religion Was Prophetic

The black Spirituals were a prophetic response to a crisis predicament. They had both an existential and eschatological dimension. They described in forceful language the slave's dread existence and also pointed to an ultimate arrangement where justice would prevail. In the midst of slavery, two separate and distinct views of God and man emerged. The master saw slavery as a natural relationship between the white man and the black man, a relationship sanctioned by a God who expected the white man to civilize and Christianize the black man. The slave saw slavery as an abomination which devastated his life and culture and awaited a God who would bring slavery to an end either by enlightening the conscience of the white man, or by assisting the black man in his revolt against it. For the slaves, Christianity created solidarity among a people who lacked social cohesion and a structured social life. Religion was part and parcel of the revolutionary mentality of those oppressed people. The charismatic Methodist and Baptist preachers offered a fiery message of salvation and the prospect of escape from their earthly woes, and their emphasis on "feelings" found a ready response in the slaves who were repressed in so many ways. The slaves, torn from their homeland and kinsmen, their cultural heritage lost, were an isolated and broken people. In the emotionalism of the camp meetings and revivals some social solidarity and psychological exhilaration, even if temporary, was achieved. This growing importance of Christianity as a social, adhesive force among the slaves alarmed the slave holders, who remained on guard against African religious practices which (although played out in a Christian context) could provide an opportunity for revolts. The original strategy of the white slave

51. Joseph Washington, Jr., *Black Religion* (Boston: Beacon Press, 1964), p. 33.

holders of introducing the slaves to Christianity in the hope that this would give them (the slave holders) more control, turned out to be counterproductive. Acquainting the slaves with the Bible gave to them themes of freedom and deliverance that their own tribal stories did not. The white population was, as it were, unconsciously providing the setting and tools for black uprisings and violence. The Bible was the means by which the blacks acquired a new theology. The white slave owners went through a series of perceptions on the question of Christianizing the slaves. On the one hand, they felt this would make the slaves more manageable but, on the other hand, there was an uneasy awareness that the claim to the same Christian fellowship would, eventually, create the concept of spiritual equality between master and slave. Peter Kalm, a Swede travelling in America wrote that the plantation owners are "very ill pleased" that Christianized blacks would consider themselves on the same level as their masters.[52] The missionaries themselves found so many obstacles to the conversion of blacks on the part of whites that they often hid their true motives in wanting to baptize. Winthrop Jordan points out that the white clergy who served the black communities, in order to maintain their freedom to preach, made the case that baptism and the propagation of the gospel were attractive devices for slave control.[53] And later, when blacks became more numerous in certain churches than whites, and wished to participate more fully in the liturgy, there was a great fear among the whites that some of the blacks might even take leadership roles. In Virginia, Baptist Church regulations stipulated that persons of color should not be allowed to preach, under pain of excommunication. This led to the development of the first separate black church in the South, the Baptist Church founded in 1773, at Silver Bluff, South Carolina, by a slave, David George, his wife and five other slaves. David took over the church while the area was still under British occupation, but fled to Nova Scotia when American forces reclaimed their area. The late 1700's and the early 1800s saw the founding of other all-black churches. There followed the founding of the First African Baptist Church of Savannah and, in 1803, a Second African Church organized by members of the First. A few years later, a Third came into being.

---

52. Peter Kalm, *Travels Into North America*, in Vol. 13 of *A General Collection of the Best and Most Interesting Voyages and Travels*, edited by John Pinkerton (London, 1812), p. 503.
53. Winthrop Jordan, *White Over Black: American Attitudes Toward the Negro, 1550-1812* (Baltimore, Md: Penguin Books, 1969), p. 191.

The Black Preacher

These African Baptist churches were led by black pastors and came into being by separating from congregations that were dominated by whites. Once again, we observe that the origins of the black church were in the context of separation and protest. The churches of blacks had no priesthood to serve them, only preachers. But, since all forms of organized social effort were forbidden among the slaves and in the absence of an established priesthood, the black preacher played the most important role in the church of the slaves. The black preacher was "called" to his office and through his personal qualities achieved a position of dominance. The "call" was supposed to have come through some religious experience. One qualification which the Negro preacher among the slaves needed to possess was some knowledge of the Bible; another qualification, the ability to sing. From the beginning of religious expression among the slaves, "shout songs" preaching on the part of the leader was important. This preaching consisted of singing sacred songs, the Spirituals. Since the slaves saw in the Old Testament Scriptures the symbols of their salvation, the preacher who was erudite enough to explain the Scriptures with credibility was especially valuable. Many black preachers of this kind waxed eloquently on the contributions made to the white as well as the black world and culture by Egypt and Ethiopia. They continually reminded their congregations of the glories and heritage of the African past, thereby boosting their morale. Present-day black scholars maintain that white scholars, and more specifically the so-called "American School of Anthropology," engaged in historical revisionism by making the case that the Hamitic people were white, attempting to negate the historical reality of which the black man was very proud.

Some of the great black preachers who expounded on the relationship between the black man in America and the Egyptians were Richard Allen, Lott Carey, Henry Highland Garnet, Alexander Crummell, Weward W. Blyden and Bishop M. Turner. Bishop James Hood treated the subject with fervor, commentating on Psalm 68:31:

> But the promise is that princes shall come out of Egypt, and that Ethiopia shall soon stretch forth her hands unto God. That this prophecy is now in the course of fulfillment the Negro Church stands forth as unquestionable evidence...[54]

---

54. Wilmore, *Black Religion and Black Radicalism*, p. 121.

While white slave masters felt comparatively safe having their slaves worship in the same churches as they did and under their supervision, the "invisible institution" was coming into being at night, in the bush and in other forms of meetings and gatherings. It was the black preacher who formed the theology of the slaves' Christianity. In the absence of an established priesthood, the black preacher, a fellow slave, had to work out the details of what would constitute a "church." Prudence in a preacher was very important. It was necessary that he be a man who articulated the desire for freedom, but did so with prudence. The congregation looked for a man accustomed to subtlety, and practiced in that form of deceit that enabled the slave to preserve self-identity in an atmosphere designed to demean whatever was left of his proud African heritage, music and worship.

The African slave paid polite attention to the Master but he never did submit to the Master's concepts of God or any version of the Christian gospel that institutionalized his oppression. The first religious leaders he recognized were those men who had either learned their craft in Africa or were taught by someone who had. Elements within African religion that could be used for resistance to acculturation did not disappear so much as they took on more and more subtle forms. There was always resistance to the total subsumption of everything black under everything white. The white preacher tried to counter this resistance. The implications of a freedom in Christ as discovered in the New Testament and Reformation theology was often studiously avoided by those few white missionaries who tried their hand at ministering to the slaves. They preferred: Ephesians 6:5, "Servants be obedient to them that are your masters according to the flesh, with fear and trembling, in singleness of your heart, as unto Christ."

In Haiti, the African brothers and sisters of the American slaves found in their African rites and the priests whom they carried into the mountains with them, the source of their fierce resistance to the colonial power. Their leader, Toussaint L'Ouverture, a Roman Catholic, was also, before his guerilla activity, a medicine man, or root doctor. Their priests saw to it that the white man would never be able to destroy that self-identity. The preacher had to be dedicated to freedom but prudent enough to preserve the lives of his congregation from the sometimes deadly wrath of the white establishment.

The Black Preacher Develops a Mature Religion

The black preacher developed a religion that was a mixture of Christian belief and the secularizing tendencies that went with a desire for political freedom. It was a Christianity of protest. It was a religion that contained amazing joy and happiness, as blacks became convinced that the Lord and Jesus were on their side and would eventually free them from their bondage. It was a religion that was substantive but not comfortable, that spiritually buttressed a war with the white man, a war in which one side had all the physical armaments and the other, the power of prayer and the assistance of a just God. Booker T. Washington learned all about this war while praying:

> ...the first knowledge that I got of the fact that we were slaves was early one morning before day, when I was awakened by my mother kneeling over her children and fervently praying that Lincoln and his armies might be successful...[55]

The black preacher had done his job very well. He both passed on an ancient tradition with his words and his songs and prepared the hearts and the minds of his flock for whatever had to be done, actively or passively, for freedom. In referring to the religious songs sung on the plantation where his parents were slaves, Washington says:

> Most of the verses of the plantation songs had some reference to freedom... they gradually threw off the mask, and were not afraid to let it be known that the "freedom" in their songs meant freedom of the body in this world.[56]

The black preacher also prepared the ways for what was to become the theology of the "suffering servant." He taught his people that God had chosen them to save the souls of the white man, that slavery was a sign of having been chosen to help God in His task of saving the white Christians' souls. Washington speaks of what he had learned about the morality of the white man:

> The wrong to the Negro is temporary, but to the morals of the white man the injury is permanent...The white man who begins to break the law by lynching a Negro soon yields to the temptation to lynch a white man.[57]

---

55. Booker T. Washington, *Up From Slavery* (New York: Airmon Books, 1967 edition), p. 18.
56. Booker T. Washington, p. 25.

Washington continued, however, to articulate a religious feeling that he did not get from the black preacher. He said that the work which had been done by all denominations, white included, was Christ-like in its treatment of black men. His fellow black preachers disagreed. They were prepared to say that the white man might be converted as a result of the suffering of the black man, but they were not prepared to admit that the white church had done anything of good for the black man. He explains their reaction to his attitude:

> Dr. Lyman Abbott, editor of the Outlook asked me to write a letter for his paper...on the exact condition...of coloured ministers in the South. What I said soon reached every Negro Minister in the country...I think every association of my race, that met, did not fail to pass a resolution condemning me...[58]

The implied comparison between black and white ministers was, as Washington indicated, most offensive to the ears of the black community which, in its attitude of protest against the way they had been treated, would never admit to any inferiority on the part of the black man, especially in the matters pertaining to Christianity or its exercise. Perhaps one could argue that, in the main, the use of evangelism as an instrument of socio-economic improvement in the black community failed to attain its goal. Ghetto housing, racism and unemployment was then, and continues to be, pervasive.[59]One could also argue, however, that things might have been much worse.

The Black Church

The ambivalence toward the morality of slavery on the part of the white church hastened the formation of the black church. For example, the Presbytery of Transylvania in 1797 hesitated:

> The question was put to the presbyter: "Is slavery a moral evil?" The vote gave the answer as "Yes." A second question was put: "Are all persons who hold slaves guilty of a moral evil?" The answer given was "No." When a third question attempted to get the presbytery to decide, if not all slave holders, which of them should be considered guilty of a moral evil, the answer was: "resolved that the question...be put off until a future

57. Ibid, p. 105.
58. Ibid, p. 141.
59. James H. Harris, *Black Ministers and Laity in the Urban Church: An Analysis of Political and Social Expectation (University Press of America, Lanham, Md.) 1987, p. 51.*

day." The Presbyterian historian Andrew E. Murray remarks dryly: "This day seems never to have arrived."[60]

The cold climate of the white church encouraged the black community to worship on its own, in the warmth of their own togetherness. For the more than four hundred years of the black experience in the Americas, eighty-five percent of all black Christians have worshiped with people of their own race in all-black congregations. Black religion has existed in Canada, the United States, the Caribbean, Central and South America, and Africa for hundreds of years, sometimes in the form of orthodox Christianity and sometimes in forms antagonistic to Christianity, especially in the rites of new-African cults and sects. The African Methodists began with a revolt against the control of white presbyters. Three of the earliest black Baptist preachers (David George, George Lisle and Amos Williams) fled slavery and founded new congregations in Nova Scotia, Jamaica, and the Bahamas. The black church was the core of the secular as well as the religious movement to bring about the total abolition of the condition of slavery.

Black Christianity, as expressed in the Black church, can be said to be the same and yet different from the Christianity practiced by white people. It is different in that it reflects the historic struggle against racism and oppression that black people have experienced, sometimes in the name of Gospel values. It differs in its music, its style of preaching, its ethical commitments and ideas about social justice. It is strongly communal and political in its orientation. And always there was the concept of "freedom." The Exodus of the Old Testament dominated the New Testament Christianity of the black church. James Smylie writes about the strong liberation themes of Martin Luther King, Jr. explaining that the Exodus, for King, was an archetypal experience whose language provided him with the metaphorical language he so skillfully used to articulate the experience of being black in America.[61] Because "freedom" was a core concept in the evolution of the black church, because it implies political action, the line between the secular and the sacred, so precisely drawn in the white church, does not exist in the black church. A basic characteristic of the black church is the unity of secular and the sacred, the profane and

60. Gayraud S. Wilmore, *Black and Presbyterian* (Philadelphia: The Geneva Press, 1983), pp. 62-63.
61. James Smylie, "On Jesus, Pharaohs, and the Chosen People: Martin Luther King as Biblical Interpreter and Humanist," *Interpretation 24* (Jan. 1970), p. 81.

the holy, everyday life and the passion of religious experience. The white church's habit of dividing reality up into two parts, was rejected by the black church; this would have been tantamount to accepting the extinction of the race because political agitation was necessary for survival. The black community, for reasons of its survival, has rejected the polarity of the holy and worldly aspects of life. The black community is a people who have had to look at the hard realities of their existence, a people who had to be religious while dealing with the existential realities of their pitiful condition. Black religion was not meant to be a fourth American faith, but is the result of the peculiar sufferings and trials, hopes and victories of the American black.

Nat Turner

Black uprisings brought severe reactions from plantation owners who urged legislatures to ban religious meetings of blacks. It was widely reported that Nat Turner, like Denmark Vesey, claimed religious sanction for his actions. This association of slave rebellion and slave religion brought about a wave of protests and proscriptive laws throughout the South. Most of laws banned religious services conducted by black preachers. Black preachers such as Denmark Vesey and Nat Turner actively fought slavery in their own way. Their churches served as meeting places where protest strategy was planned, and as way stations on the underground railway. Turner, a Baptist preacher and slave, not only preached rebellion against white slave owners, but became the leader of the most serious and dramatic slave revolt of his time. This revolt took place in Southhampton, Virginia, in the summer of 1831, showing that the slaves' religious tradition could, on certain occasions, provide the seed-bed for armed rebellion. In the years before the rebellion, the mystical Turner, who saw himself as a divine instrument to help deliver his race from bondage, claimed to have had innumerable visions, one of which he described as white spirits and black spirits engaged in battle. The sun darkened and blood flowed in streams. He was convinced that God had instructed him to rise up and prepare himself to slay the enemies of his race. He stated explicitly that his intention was to carry terror and devastation wherever he and his followers went. On the morning of August 22, 1831, armed with an axe, Turner entered his master's bedroom and slew him, his wife, and three others in the household. He then, together with his followers, roamed the countryside and, within a matter of hours, systematically massacred whites. At the home of his slave master whom he had slain, Turner and his accomplices dressed like infantrymen as they set out to enact the bloodiest slave insurrection in American history. On Sunday, October 30, he was apprehended and on November 11, 1831 he was hanged. Before he was sentenced, Nat Turner, in response to the Court, said that he had nothing to say. He made no appeal for mercy and showed no sign of repentance. Fifty three blacks were arrested in connection with the revolt and twenty-one of them were acquitted. Twelve slaves were moved out of the state and twenty were hanged. More than one hundred slaves were killed before the revolt was crushed.

Nat Turner's revolt sent shocks of fear throughout the white community, which became very suspicious of every black preacher and every black congregation. Herbert Aptheker reports the reaction of Governor Floyd of Virginia:

> From all that has come to my knowledge during and since this affair, I am fully convinced that every black preacher in the whole country east of the Blue Ridge, was in on the secret.[62]

DuBois reports what also happened:

> A wave of legislation passed over the South prohibiting the slaves from learning to read and write, forbidding Negroes to preach, and interfering with Negro religious meetings...[63]

One of the results of Nat Turner's revolt was to hasten the development of the black church. The reaction of the white community, stifling the religious expression of the black community, boomeranged as blacks became more determined than ever to form their own churches.

Denmark Vesey

I should make mention of the other major revolt planned but never executed. This revolt was planned by Denmark Vesey, a young intellectual with an absorbing interest in black religion. After receiving his freedom from a Captain Joseph Vesey in 1800, Denmark settled down in Charleston, South Carolina, working as a carpenter. For twenty years he spent a

62. Herbert Aptheker, *American Negro Slave Revolts* (New York: International Publishers, 1943), p. 305.
63. W. E. Du Bois, *The Negro Church* (Atlanta University Press, 1903), pp. 25-26.

lot of time with the community of free blacks that had grown up in Charleston and was considered a community leader. A study of the Old Testament led him to the conclusion that his own people resembled those captive peoples who yearned to reach the Promised Land. Joshua and his plans for the battle of Jericho particularly fascinated and then inspired him. Vesey worked hard and organized two to three thousand blacks to move when given the signal. The date for the revolt was set as July 14, 1822. One slave, Devany, was so excited about the plan that he told his master, who notified the authorities and the revolt never came to fruition. The trial notes claim that Denmark Vesey used Scriptural texts to win supporters for the insurrection. His organizational ability is reflected in the fact that three to nine thousand slaves had been mobilized. Monday Gell, one of Vesey's lieutenants, even wrote to President Boyer, of Haiti, asking for assistance. But all this planning came to nought and Vesey and his closest collaborators were executed on July 2, 1822. Vesey's planned revolt failed but his determination to do something to free those of his race still in slavery, inspired revolts, conspiracies and incidents of burnings throughout the South. Here also it was noted that the planned revolts took place in the context of Old Testament study and organized religious meetings in churches throughout the South. The black church had become a revolutionary force to be dealt with.

The Church—Focus of Protest

The church continued to remain the major forum where slavery was discussed. Slave revolts and uprisings, successful or not, served both to gain more respect for slaves on the part of their white masters and emboldened the slaves themselves to articulate their demands. Albert J. Raboteau notes:

> As early as 1774, American slaves were declaring publicly and politically that they thought Christianity and slavery were incompatible. In that year the governor of Massachusetts received "The Petition of a Great Number of Blacks of This Province who by divine permission are held in a state of slavery within the bowels of a free and Christian Country"... How can the master be said to Beare my Bordon when he Beares me down with the Have chanes of slavery and operson against my will..."[64]

64. Raboteau, *Slave Religion*, pp. 290-291.

Part of the protest inherent in black religion expressed itself in the denial of the structure of white morality. It was a virtue to steal from whites, to lie to them, to be deceitful to those men and women who enslaved them. Those engaged in pastoral work in the black community can testify to the great hesitation found there in condemning acts of violence on behalf of members of the black community, at least in the presence of whites. The acts perpetrated are either denied outright or ascribed to the denial of civil rights and the opportunity for employment. That same hesitation can be seen in black communities today, in the Crown Heights section of Brooklyn, as blacks called for the release of a young man accused of stabbing a Jewish person. In Washington, D.C. Whites were surprised and shocked when former Mayor Barry, of Washington was re-elected to another, though lesser, political office. The White perception is that blacks have no sense of law and order, no self-discipline. The reality is that blacks have been the object of "law and order" from the time of slavery, have been hung by "juries" all over this land. What can we in the white community expect now? Where were we when the hangings were taking place? Why were we not shouting then? All of us, black and white, agonize over this terrible curse of racism, but we cannot ask for "reason" until we become reasonable.

The Jesus who walks with members of the black church remains the resurrected Lord. When the Black community sings "I Want Jesus to Walk with Me," they are calling on Jesus both to cleanse their souls and to be their political Redeemer. As most black leaders say, the black community has come a long way and it has a long way to go, and if Jesus will continue to walk with them, they will ultimately triumph. The hope for freedom, therefore, continues to rest in the black church and the articulation of this hope remains with the black preacher. It is no accident that the same Martin Luther King Jr. who led public demonstrations, also shepherded a flock in his modest church in Atlanta. It is no accident that Jesse Jackson, a black preacher, became a Presidential candidate. Protest and Political activism remain essential components of the black church. The black community will support black politicians, but on their pilgrimage to freedom they put their trust in the black preacher and the black church.

The Black Church Becomes Distinct

Although the "invisible institution" provided space for worship that was heavily influenced by African retentions, the black church had remained pretty much an underground church. The first black congregation of record was formed in 1758 near what is now Mecklenburg, Virginia. The Joy Street African Baptist Church of Boston dates from 1805, while the famous Abyssinian Baptist Church in Harlem was founded in 1809. As more and more blacks, now free from the domination of white slave masters, fled white churches, preachers arose to form them into their own independent churches. Richard Allen, who became well known for starting an independent Methodist Church in Philadelphia, began even earlier to gather blacks without regard to denomination for prayer and reflection on African culture. These groupings eventually became known as free African societies and the style of these meetings became the dominant style of the developing black church. Gayraud Wilmore describes them:

> The free African societies did not express the need for cultural unity and solidarity only...it was usually their objective to provide not only for religious needs, but for social service, mutual aid, and solidarity among "people of African descent."...They created, therefore, the classic pattern ....autonomous worship in the Afro-American tradition, and the solidarity and social welfare of the black community.[65]

As this black church development was taking place, its members in the North, especially those of the African Methodist Episcopal (the church founded by Richard Allen) took on the task of assisting their brothers in the South who were still living under the yoke of slavery. The smaller congregations of the AME Church served as stations of the Underground Railroad. All the best-known blacks associated with the underground railroad were members of this church. To the credit of the black church, once it had gained its own independence and acceptance as part of the religious life of America, it continued first to assist those blacks still in slavery and later, up to our own day, to defend the rights of the oppressed and downtrodden. It widened its vision to the African continent and more recently to Latin America. The black church was born in protest and protest remains an integral part of what it is.

Summary

We have seen that black religion in America has been a critical reflection about God and religious faith from the perspective of slavery, racial oppression and African cultural adaptation. The theology which devel-

65. Wilmore, *Black Religion and Black Radicalism*, pp. 82-83.

oped in such a milieu was necessarily characterized by this reflection, bearing within itself the traumatic effects of the black social condition and expressing itself in the rhetoric of protest. The black theologian will argue that the contradictions found in the white American church are derived from its historic connection with slavery and the prejudice of color. He will argue that the problem of white racism, its disestablishment and dismantling, is the task of black theology. Black theology stands in opposition to the continued domination of "black" by "white." It seeks a true brotherhood of man based on the Gospel message, and freed from the idealization of the Caucasian man. It is, therefore, a revolutionary church, for it seeks to destroy an old condition and replace it with a new condition, a blessed condition in which there is equality of opportunity for the black person in America. The black minister does not have the luxury of simply baptizing and confirming; he must also strategize. In his church the political strategies are conceived and executed. In his church the wherewithal for the black community to survive, to be clothed, to eat and to sing with joy, must be found. There is no other "place" for the black man in America. If the church does not politicize, revolutionize, criticize, then the black man would simply have to lay down and die. In Chapter Three we shall examine the sources and content of the black theology that both drives and is driven by this black church. It is an academic discipline inherited from the antebellum black church meeting. It must now "give back" by providing enlightened leadership, leadership that is wise, subtle and strong.

CHAPTER THREE

SOURCES OF BLACK THEOLOGY

As we have seen, Black religion was transformed in the milieu of slavery as protest against both the conditions of slavery and the white Christianity that accepted these conditions. Since black theology has evolved from black religion and the conditions which made black religion a religion of protest, black theology is itself a theology of protest, a protest against the nature of human existence as experienced by the black community. It has developed, therefore, as a critical corrective of the content of classical theology. Noel Erskine comments:

> Black religion began the process of decolonizing theology when it insisted that God was the freeing one who was at work in history setting the victims free.[66]

And:

> Colonial theology also explicated a false eschatology, which was designed to postpone freedom to the afterlife...But black people, in the expression of black religion, pressed for historical freedom.[67]

There was, even in colonial America, prophetic abolitionist voices that attempted to influence the course of events. David Swift observes:

> Drawing on the same Enlightenment stress on universal human characteristic, and rights as Jefferson had, these abolitionists endorsed monogenesis and attributed racial differences to variety in environmental influences exerted over long periods of time. Activist black clergy strongly supported this position, setting the biblical account of the Creation in the book of Genesis alongside the Declaration of Independence as complimentary authorities.[68]

---

66. Noel Leo Erskine, *Decolonizing Theology* (Maryknoll, New York: Orbis Books, 1981), p. 116.
67. Ibid, p. 117.

The World View of Black Theology

The theological world view of black theology is that of a prophetic Christianity, nuanced by the cultural outlook of Afro-American humanism. Black theology is neither escapist in content nor a sophisticated political ideology. It is, however, pregnant with the deeply tragic quality of the everyday life of a culturally degraded, politically oppressed and racially slurred people. Cornel West theorizes that the radically comic character of Afro-American life flows "primarily from the Afro-American preoccupation with tragedy, a preoccupation significantly colored by the Black Christian world view."[69] George Kelsey, of the Drew University Theological School, accuses white Christians of having two faiths, Christianity and racism. He contends that racism is complete as a conscious, rational system of belief, including apologetics. A primary source of black theology, then, is the need to explain the real meaning of Christianity to whites, and this explanation thus becomes part of the content. There is a dynamic relationship between source and content in black theology because, unlike modern Biblical scholarship, the source is always a black experience and not an archeological or academic discovery.

Black theology is future-oriented. The hope of freedom and salvation found in the Gospel is the future to which it aspires and seeks to make a significant contribution. It is, therefore, a dynamic theology which seeks to explain the mind of God regarding the most serious moral issues in America today. If this is its strength, it also defines part of its limitations. Black theology admits that in its preoccupation with these issues, it does neglect to emphasize traditional theological issues. There is no apology for this because the American moral issues to which it addresses itself have theological underpinnings that need to be uncovered and dealt with. Major Jones explains:

> Black theology differs from traditional theology by the simple reason that it may not be as concerned to describe traditional themes, as the external nature of God's existence, as it is to explore the impermanent, paradoxical, and problematic nature of human existence.[70]

---

68. David E. Swift, *Activist Clergy Before the Civil War*, (Louisiana State University Press, Baton Rouge)1990, p. 26.
69. Cornel West, Prophesy Deliverance (Philadelphia: The Westminster Press, 1982), p. 151.
70. Major Jones, Black Awareness (Nashville: Abingdon, 1971), p. 14.

The Emergence of Black Theology

The deaths of Denmark Vesey and Nat Turner had a chilling effect on the outward expression of religion as protest. Pervasive in the black community was the feeling that the overwhelming strength and power of the white community spelled death and destruction to those blacks who articulated their pain and frustration at their condition of slavery and poverty. Freedom from slavery was not the liberating experience that the slaves had anticipated. In the place of physical slavery, white America imposed a code of social and economic behavior which denied blacks the ability to enter meaningfully into the economic and political life of the nation. The growing realization that freedom from slavery was not a signal for total emancipation led the intellectual leaders of the black community to withdraw into themselves and to reflect on the meaning of their new condition, its limitations and its possibilities. And since the intellectual leadership of the black community remained, for the most part, the black preacher and pastor, their reflection on the new condition of the black man became the dominant reflection of the black community. Black perceptions at that time came, primarily, from the black church. The black preacher spoke on behalf of a physically free but muzzled community.

Dubois' Contribution

The atmosphere of the post-bellum period is described by DuBois, in part, by sketching the young Southern Negro who is tempted, against his instincts, to remain silent and wary, flattering and pleasant, hiding his real thoughts. Dubois's *The Philadelphia Negro* (1896), a seminal work, was the first empirical and sociological analysis of Black Americans. In the same year he approached the U.S. Commissioner of Labor, Carroll D. Wright, with a proposal to study the black population in a small Virginia town and, in 1898, speaking before the American Academy for Political and Social Science, he said that the University was the proper agent for Black Studies. After a promising beginning at the University of Pennsylvania, progress was slow, so he moved to Atlanta University, which published a series of 18 studies on Black America.

We observe in DuBois' writings the connection between true freedom and religious practice. He does not encourage, as did many militants of his time, an attack on religion as one of the major contributors to oppression, an inhibitor of true liberation. That never happened in the black community. Major revolutions in the white community have often led to an attack on religion. In the black community, however, it is religion that has led to revolution and the articulation of protest and grievances about economic and social restraints and injustices. DuBois' writings, as an early source of black theology, then, encourage faith, fidelity to the gospel, and belief in Jesus as the Messiah, the Liberator of the oppressed. Black theology's roots are revolutionary, while remaining bible-centered.

Back to Africa Movement

The first theological response to the new condition of neo-slavery was the determination that for black folk to save their soul and be spiritually as well as physically freed, they must return to Africa, where they could be themselves, where they could drink deep once again of the religion of their ancestors. The process of repatriation back to Africa began even from the earliest beginnings of slavery as those few slaves who managed to be free returned to their homeland form South America, the Caribbean, and, later, from North America. Gayraud Wilmore quotes the Reverend Rufus L. Perry on this matter as observing that the ancient Egyptians, Ethiopians and Libyans were the ancestors of the race of Ham, the Negro race. Perry was an ex-slave and the pastor of a Baptist church in Brooklyn. By referring to the "Egyptians," "Ethiopians" and "Libyans" he was making a connection between ancient, proud civilizations produced on the African Continent and the poor black slave on a Southern plantation or in a Brooklyn ghetto. If the black man was to be encouraged to return to Africa, he must be made to feel that he is going back to something of which he can be proud.

The Back-to-Africa Movement also encouraged a missionary effort on the part of American blacks to Christianize their brothers and sisters "back home." Pan-Africanism, black nationalism and the establishment of the black church was really one Movement with many parts and energized by religious feelings and religious leadership. Marcus Garvey, a brilliant and bold Jamaican, was one of its leading advocates. It was he who first introduced the concept, now very familiar in black theology, of the black God. Kelly Miller, a writer for the *Amsterdam News,* expressed his misgivings about Garvey's views:

> Marcus Garvey some little while ago shocked the spiritual sensibilities of the religious world suggesting that the Negro should paint God black.[71]

Many writers and preachers, black as well as white, scoffed at Marcus Garvey, but E. Franklin Frazier noted:

> The intellectual can laugh if he will, but let him not forget the pragmatic value of such a symbol among the type of people Garvey was dealing with.[72]

Garvey, head of the Universal Negro Improvement Association and the "Provisional General of Africa," got into trouble with the law with a conviction for the fraudulent use of the U.S. mails, during the course of selling stock in the Black Star Line. It was this Line which was to transport negroes back to Mother Africa. He began a five-year jail term in 1925, but the sentence was commuted by President Coolidge, upon the advice of Attorney General John Garibaldi Sargent. Since he had never taken out final citizenship papers, he was eligible for deportation as an undesirable alien. He made his farewell speech from the top deck of the S.S. Saramacca, sailing from New Orleans to Panama, from where he made his way to Jamaica. "I leave America fully as happy as when I came," he said, explaining that his relationship with the Negro people was a pleasant and inspiring one, and that he would always work on their behalf.

The Back-to-Africa Movement received a severe set-back and lost the support of most black preachers and pastors when Samuel Mills and Ebenezer Burgess, white agents of the American Colonization Society, themselves began to advocate that blacks return to Africa. Southern Whites saw the Movement as a means of ridding the South of blacks from the South, whom they feared would eventually threatening white advocacy of the Back-to-Africa Movement killed the Movement, and the theological focus shifted to black separatism. Edward W. Blyden, a Presbyterian, Theodore Holly, an Episcopal bishop, and Henry M. Turner of the AME Church were amongst its strongest advocates. Turner, also a bishop, was especially important to and made a strong impact upon the young black clergy, in his advocacy of a rising black consciousness that would separate blacks from the concepts of the religion of the white

71. Kelly Miller in an article in *Amsterdam News,* 16 Feb., 1927.
72. Franklin Frazier, "The Garvey Movement," *Opportunity,* IV (November 1926), p. 147.

church. Turner was a major critic of Booker T. Washington, whose willingness to compromise with the white community Turner considered tantamount to disloyalty to the black race. So strong were Turner's views that he became isolated in his own religious denomination. He preached that black religion rests upon a constant remembrance of the centuries of oppression suffered by black people, and not on the spiritual offerings of the white church, modified by the black preacher. The post-bellum attitudes of the black church that men like Henry Turner fought against is described by James Cone:

> Before the rise of black theology, black churches accepted uncritically the theology of white churches, using their doctrines and creeds as if the racist behavior of whites had no impact upon their views.

Black Church Defended

Not everyone agreed with this evaluation of the black church. Some saw this as too harsh a criticism, preferring to believe that any "acceptance" of white theology on the part of the black church had much more to do with survival than with conviction. Gayraud Wilmore describes the reaction of the black church to the charge that it was not "black" enough, saying that the term "black theology" emerged as a reaction to the allegation that the black church was devoid of theology.

The ante-bellum black religion of protest never did die in the black community, but it would be accurate to say that its protest became less strident and articulated by fewer preachers and intellectuals. One such intellectual was Martin Delany, a physician and journalist with theological skills who wrote in the mid-1800s. Delany believed that it was foolish of blacks to delude themselves into thinking that their condition would be relieved by prayer and self-discipline. He criticized the church for teaching an excessive amount of otherworldliness, and preached of God the Liberator. In this, Delany was giving theological formulation to the underlying themes of the black Spirituals which, sung under conditions of slavery and close scrutiny by white slave masters, were characterized by subtle and clandestine meanings. Delany was convinced that the hope for black people lay in self-help and a refusal to allow white people to theologize on their behalf. Delany's attempt to relate an exegesis of Scripture and social conditions was an early articulation of black theology.[73]

The most articulate black preachers and pastors understood liberation as being a function of the Gospel message. The black preacher became an expert at balancing the need to protest the black condition on the one hand and calming white fears of black violence, on the other. These two contrasting tendencies, toward radicalism and compromise, represented two strands of a survival tradition. It was the ever-present threat to the security and peace of black existence coming from the white community, and the inability of the black community to respond in terms of power, that led to the apparent "cooperation" that James Cone criticizes. Cone, however, in later writings has given much more credit to the difficulties with which black preachers and their congregations had to grapple in the pre-civil-rights atmosphere. He is, today, much more sympathetic to the early black theology which mightily held on to the desire for freedom, white accommodating, where necessary, for survival. Paul Laurence Dunbar, in his poem, *We Wore the Mask*, is eloquent as he describes the psychological dexterity characteristic of the black man and woman. "We wear the masks that grins the lies...With torn and bleeding hearts we smile."[74]

Theological Adjustments

The Black denominations adopted the doctrines of the white church, but blacks took these doctrines and made them their own in selection, just as they were doing with the Old Testament. Passages from Scripture that commended obedience and submission were substituted by passages that proclaimed the freedom of the children of God, and announced the coming and function of the Messiah. This process of selection and refinement of biblical passages for purposes of motivating the black community took place in the South during the Reconstruction period. The theological base for activism also developed during this same period, as the black church led the black community in the domain of political involvement. It sent a preacher, the Reverend Hiram Revels, as its first Black United States Senator from Mississippi. While white theology continued in its determination to be otherworldly, the black church during Reconstruction preached the necessity of political activism. Since the roots of black theology lie in protest, it is not surprising that political activism has always been an inte-

---

73. Martin Delany, *The Condition, Elevation, Emigration and Destiny of the Colored People of the United States, Politically Considered* (Philadelphia, 1852), p. 38.
74. Paul Laurence Dunbar, "We Wear the Mask," *International Library of Afro-American Life and History: An Introduction to Black Literature in America*, ed. Lindsay Patterson (Cornwells Heights, Pa.: The Publishers Agency, 1976), p. 95.

gral part of black theology. From 1870 to 1901, twenty blacks served in the Senate. A reporter described the first South Carolina legislature of the Reconstruction era:

> The Speaker is black, the Clerk is black, the doorkeepers are black, the little Pages are black, the Chairman of the Ways and Means is black and the Chaplain is coalblack...these men...have been themselves slaves...[75]

The preacher and pastor who urged his black congregation to get involved in politics was making the statement that political activity, for the black man, was religious activity. The theological agenda of the black church was being formed in the early beginning of the struggle for civil rights. The black preacher did not reject the unworldly element in the Christian tradition, since the resurrection proclaimed the immortality of the soul and eternity with Christ, but he did add the dimension of actively seeking, in the field of politics, the freedom and dignity proclaimed in the pulpit. James D. Tyms quotes O. Clay Maxwell, former president of the National Baptist Sunday School and Training Union:

> The greatest weakness of the religion of our time, and times past, is that it draws circles around the areas of life where the religion of Jesus belongs and shuts him out of vital areas of human interest and activity.[76]

Such a statement has theological implications which mirror the this-worldly emphasis of redemption in the Old Testament, implications which black theologians claim have been neglected by classical theologians, or at least subordinated to an emphasis on the otherworldly. At the end of the Reconstruction period (1877) the black church slowed in its growth, but it did deepen its theological study. It was a period of reflection, a time of refining its own interpretation and understanding of the Scriptures. Olin P. Moyd calls this the Maturation Period of the black church. He lists its accomplishments:

1. By the end of the of the maturation period...The Black denominations were autonomous.

2. Blacks had fully developed their modes and styles of worship...highly tempered with African survivals.

---

75. Quoted by Church Stone in his *Black Political Power in America* (New York: Exposition Press, 1965), p. 8.
76. James D. Tyms, *The Rise of Religious Education Among Negro Baptists* (New York: Exposition Press, 1965), p. 8.

3. Adaptations and reinterpretations of doctrines and Scriptures by Blacks were fully developed.[77]

"Adaptations and reinterpretations of doctrines" meant the creating of a new theology, a black theology based on the black soul and the black experience. Nor did this new theology remain in the Reconstruction South. The black church movement to the North began and quickly developed new and creative ways of expressing itself, from storefront churches to more elaborate church structures. Redemption remained its theological base. The theology of this growing black church has been truly existential in that it asserts that existence is prior to essence, and one's personal experience is prior to any theory about the world or reality. The focus on "existence" was eventually to lead, in the 1970's, to the black theologian's interest in Marxist analysis, since its interpretation of the world in anthropocentric terms was well-suited to both the needs and the mood of black theology. John Mbiti, an eminent African theologian, explains that African ontology is basically anthropocentric, man as the center of existence.

Black theology, then, views theology as a rational study of God in the light of the existential situation of an oppressed community. The forces of liberation are related to the central figure of the Gospel, the liberator, Jesus Christ. This theology ensures the black oppressed that their determination to liberate themselves is not only consistent with the Gospel, but is at the very heart of the Gospel. The black theologian does not answer his critics, who accuse him of partiality toward the black condition. For him, the universality given up on the horizontal level, appears in the deeper level of the human condition. He readily agrees that his theology appears to be un-American at times, as he challenges the economic and social structures of this society. Martin Luther King, Jr. and Malcolm X have been judged dangerous by men who perceived their writings and preaching as inimical to American society. We all remember the obsessive compulsion of FBI Chief, Edgar Hoover, who violated King's civil rights as often as King advocated civil rights for all Americans. Hoover was said to be a bible-reading man, but he could neither hear nor recognize bible messages when they came, however eloquently, from the voice of a black man.

---

77. Olin P. Moyd, *Redemption in Black Theology* (Valley Forge, Pa.,: Judson Press, 1979), p. 77.

Black Theology and Marxism

The accent on liberation themes has led to a relationship between black theology and Marxism. This relationship has been very slow in developing and its development has been characterized by great caution, on both sides. There are points of serious disagreement. Black theology claims that the basis of exploitation is not (as for Marxism) the economic difference which builds social classes. It is, rather, the racial difference rooted far more deeply in the human psyche. Black theologians, until recently, have not incorporated Marxist class analysis into their theological discourse because they have assumed that the problem of racism can be solved without a socialist transformation of the political economy. In the main, the black theologian, with his white counterpart, has associated capitalism with Christian freedom and political democracy, and Marxism with Russian Communism, atheism and political totalitarianism. However, with the rise of Latin American Liberation Theology, its endorsement of Marxist class analysis, and its strong rejection of United States Capitalism, black theologians decided to take another look. They were challenged by a form of Marxism that did not emanate from the Soviet Union, or from any other part of Eastern Europe. They were not a little surprised that this Marxist-influenced theology emanated from the most "Catholic" part of the world, Catholicism being associated, as it was, with a strong, even virulent anti-Communism. Reflecting on this "Christian Marxist" challenge from Latin America, Cone remarks:

> With the rise of Latin American Liberation theology and its affirmation of Marxist class analysis and its vehement rejection of U.S. Capitalism, white and black theologians were challenged by a Marxist perspective that was not defined by Soviet Russia or its satellites...[78]

Cornel West, more than any other black theologian, stresses the importance of Marxist analysis in the theological discourse of black theology:

> Revolutionary Christian perspective and praxis bear directly on the Afro-American liberation struggle. The central political concern is twofold: to weaken the hegemony of liberalism over the Afro-American community...and to break the stronghold of Leninism over Afro-American Marxists.[79]

---

78. James Cone, For My People, p. 177.
79. Cone, p. 181.

Here, West refers to the cautioning that emanates from the white liberal community when they deal with black intellectuals, and to the oppressive tendencies of those Marxists who tried to convince Black Marxists that the Soviet model of Marxism was an ideal one. West insists that liberation and faith cannot be separated. He favors a theology which expresses itself as deeply concerned with the vast amount of poverty in the world, a poverty that sometimes exists in the midst of plenty. He urges that at the highest levels of the church there exist a mechanism for studying, critically, the roots of exploitation and oppression, suggesting that all of our Christian resource go into this effort. The practice of our faith should be allied, he says, with a political movement displaying organization, power, and social vision, led by men of impeccable integrity.[80]

Cone contends that the reason why most seminaries offer no courses on Marx, and most American theologians seldom mention him, is because the practice and theology of American churches serve as a religious justification of the existing structures of oppression. For a refutation of Marx on the part of the church to be credible, he insists, it must express the church's solidarity with those who are fighting for their liberation from oppression.

Black theologians explored the ways in which Marxism can serve as a tool in understanding the liberating forces inherent in Redemption, but never suggested that it serve as a substitute for the Christian faith or the message of the Gospels. It is when the black theologian stresses the social consequences of sins that his language mirrors that of Marxist analysis or the liberation theology of South America. Community is defined as a spirit of mutual support of, and sharing with, each other. The rugged individualist and highly personal religion associated with the political economy of Capitalism is rejected. Communalism, not Communism, is seen as necessary for liberation, and the liberation theme is at the heart of black theology. its recurring theme. Christ is the liberator and the Christian faith promises "deliverance of the captives." It promises to let the oppressed go free.[81] Unlike their white counterparts in the white theological community who embraced, in large measure, the Marxist analysis, the black theologians tip-toed around its outer perimeter.

---

80. West, p. 122.
81. J. DeOtis Roberts, *Liberation and Reconciliation: A Black Theology* (Philadelphia: Westminster Press, 1971), p. 32.

The Prophetic Tradition and Black Theology

In the winter of 1899-1900, at the University of Berlin, Adolf Harnack delivered a series of public lectures on "The Essence of Christianity" which exerted a powerful influence on a new generation of theologians. Rudolf Bultmann, in an Introduction to the 1957 edition of these lectures, calls them "a theological-historical document of the greatest importance." Lecture VI is entitled, "The Social Question." In it Harnack says:

> There can be no doubt...that if Jesus were with us today, he would side with those who are making great efforts to relieve the hard lot of the poor and procure them better conditions of life. The fallacious principle of the free play of forces, of the "live and let live" principle - a better name for it would ne the "live and let die" - is entirely opposed to the Gospel.[82]

It is interesting that such statements were being made by a famous Christian intellectual at the start of the twentieth century, in a social atmosphere that often equated poverty with Divine displeasure at the poor. Harnack's use of such terms as "solidarity" and "brotherliness," in the context in which he uses them, is rather surprising for his time. Other white theologians have made similar observations, and if their tone is not as strident and as passionate as that of James Cone and DeOtis Roberts, it can be explained by the fact that they wrote from within communities that, by and large, were reasonably comfortable and not suffering daily from the destructive psychological effects of racism and oppression.

Cornel West examines the prophetic Christian tradition in the black community and divides it, historically, into four stages. The first stage, roughly the middle of the seventeenth century to 1863, a period of black prophetic preaching, he calls, Black Theology of Liberation as Critique of Slavery.[83] During this period, the slaves interpreted the Christian message as clearly opposed to the institution of slavery. In speeches, documents and petitions to authorities, slaves insisted that God had made them equal to their masters and, in private, had no hesitation in asserting that, in fact, given the propensity of the white race to enslave the black race, the black race appeared to be morally superior to the white race. Whites, accustomed to the high ground, were traumatized by such a daring assertion.

Nat Turner, Denmark Vesey, Gabriel Prosser and David Walker were among those black Christians who expressed black theology as a critique of slavery.

---

82. Adolf Harnack, *What is Christianity?* (New York: Harper and Row, 1957).
83. West, p. 101.

The second stage of black theological development, 1864-1969, West calls "Black Theology of Liberation as Critique of Institutional Racism." During this time black Christians began to focus a critical eye on the racist structures of American institutions which kept blacks from entering the mainstream of American social and political life. Many blacks were deprived of the right to vote, were economically and socially exploited, and became subject to segregated facilities in education and other establishments. This time-frame also saw the lynching of hundreds of black people. President Woodrow Wilson refused to sign a proposed anti-lynching law in 1916. This period saw the rise of many prophetic black leaders (Marcus Garvey being the most prominent) who favored a return of black people to Africa, and the most prominent prophetic black Christian, Martin Luther King, Jr. The early beginnings of a systematic presentation of a black theological program and method occurred in this stage, with the publication of Albert Cleage's, *The Black Messiah* (1968) and James Cone's, *Black Theology and Black Power* (1969).

West's third stage, 1969-1977, witnessed the flowering of liberation theology and the black theology of liberation in the United States. He calls this period "Black Theology of Liberation as Critique of White North American Theology". The major black theologians, James Cone, Major Jones, DeOtis Roberts, Joseph Washington, Gayraud Wilmore and William Jones emerged during this stage. These black academics took up the challenge of black power being played out in the streets of America's black ghettos, and responded theologically. It was a time of great intellectual ferment and excitement in the black community, as articulate black activists and theologians analyzed the black condition and proposed remedies and strategies. The silence of white theologians was attacked by the black theologian, who questioned the authenticity of a white Christianity and a white theology which failed to offer a critique of the black/white relationship in America. The black experience on the plantation, in the streets of America's ghettos and in the black church became the primary material for, and focus of, black theology.

The fourth, and present, stage West calls "Black Theology of Liberation as Critique of U.S. Capitalism." In the present stage, the study of Marxism has become an important part of the analysis of the causes of black exploitation in America. Profit-oriented capitalism is seen as fundamentally alien to the good of Americans in general, and the black community in particular. In a statement issued by the Black Theology Project, the affluence of a few is commented on:

> Even the material good fortune of that few is poisoned by emptiness and isolation from the people's struggle without which the mission of Jesus Christ can be neither understood nor undertaken.[84]

West believes that the present stage of development of black theology remains inadequate. He argues for a new conception of black theology of liberation which, while preserving whatever is of value from the past overcomes its earlier inadequacies and faces its present challenges with courage. The task that remains includes a more systematic study of the relationships between racism, sexism, class exploitation and black oppression, and the despair which is related to unjust socio-economic structures. West advocates a serious dialogue between black theologians and Marxists, the purpose of which would be to plumb the different sources of their praxis of faith, and to arrive at mutually agreed strategies to improve the lot of the oppressed, wherever they may live and work. He laments the fact that black theologians and Marxists, while having a similar socio-economic agenda, have not learned how to relate to each other in the pursuit of common goals. In the development of this kind of cooperation, understanding and strategy, West sees the black theological perspective moving into a future fifth stage, which he calls "Black Theology of Liberation as Critique of Capitalist Civilization."

> In short, black theological reflection and action must simultaneously become more familiar with and rooted in the progressive Marxist tradition, with its staunch anti-capitalist, anti-imperialist, anti-socialist outlook...[85]

Black and White Religion

Black theology, then, remains a theology of protest. It seeks to align itself with forces and ideologies that can be interpreted as anti-white and anti-American. It does so because "white" religion is not accepted as a religion suitable for the black community; indeed, it is hostile to the black community. The ethics of white America then, logically, is also seen as not relating to the ethics of the black community, and as equally hostile to that community. So, as in the time of slavery, tension, protest and anger remain, and is best articulated today among the black theologians. The "base" of these theologians remains the black church, but this base does

---

84. *Black Theology: A Documentary History, 1966-1979*, eds., Gayraud S. Wilmore and James H. Cone (Maryknoll, New York: Orbis Books, 1979), p. 348.
85. West, p. 106.

not prevent the theologian from entering into dialogue with Marxism. The black theologian refuses to accept the principle that either the use of Marxist analysis, or the condemnation of "white religion" and its historical praxis necessarily implies the abandonment of Christianity. Rather, it insists that the "real" Christianity must include both of these on their agenda, and more besides. They state again and again the impossibility of the destruction of religious truth, no matter what the historical circumstances, no matter what the revolutionary movements. It will be interesting to see if the breakup of Soviet Communism reverses the trend toward the critique of Capitalism and the use of Marxist terminology. My guess is that the reversal has already begun and will continue, with more emphasis again on Old Testament figures and rhetoric. In any case, the black theologian sees the critique of classical theology in terms of the further progression of the present transition period toward a future society and culture which is qualitatively different from that which developed in the milieu of classical theology. Since the white church has not addressed itself to America's most serious moral problem, it isn't considered a church whose theology can possibly be valid. The black church then, by necessity, must look to itself to articulate theology, and to those whose philosophies appear to the black community as more "Christian" than that of the Christian white community. James Cone comments:

> By defining the problems of Christianity independently of the black condition, white theology becomes a theology of white oppressors...[86]

In these strong words, Cone attempts to bring to the attention of the white community that it cannot call itself "Christian" until and unless it formally denounces racism as a National evil. If the white community cannot bring itself to do this, then it does not deserve, in Cone's opinion, the title of Christian. For him there is radical incompatibility between a church that claims allegiance simultaneously to the gospel and to racism. Major Jones agrees with Cone, explaining that the white church allowed itself to become a tool of its cultural ethos, justifying slavery on the grounds of an innate intellectual inferiority. It is difficult to disagree with their conclusions.

---

86. Cone, *Liberation*, p. 31.

Gospel and Revolution

The black theologian looks for a correlation between the eschatological origins of Christianity and the revolutionary forces that seek to build a new and better future for the black man. Allan Boesak, in South Africa, has no doubt but that God has taken sides in the South African political scene, issuing a divine call to blacks everywhere to get involved in the struggle. The black theologian in South Africa reflects on the Gospel as it relates to apartheid, oppressive security laws, separate black/white facilities and "homelands," poverty and economic exploitation. The oppressor is not a "Marxist" or a "revolutionary," but a white Christian who bases his racism and oppression on the Gospel. The black theologian's language is important. Robin Scroggs, Professor of New Testament Studies at Chicago Theological Seminary, cautions that the use of theological language outside the context of social realities, encourages a theological dialogue that takes place in a vacuum.[87] Black theology prides itself in being faithful to the Scriptures. One can scarcely sing, preach or theologically discourse in a black church without a heavy concentration of Scripture. For the black theologian the Scriptures are the voice of God and of the Gospel, not the history or pronouncements of white Christianity. That the revelation of God is found in Christ is of the greatest importance to the black theologian, for it is in the Bible that he finds that God is the God of liberation, and that Jesus Christ is God-become-man. There is no church of Christ, therefore, which does not preach liberation. The Christianity of classical theology calls for humility in the face of oppression; black theology challenges this posture. Classical theology calls for self-discipline and personal mortification and sanctification. Black theology disagrees, stating that the black race has been disciplined enough, and mortification is its daily bread. It affirms what the Gospel affirms: Jesus Christ is the life of the world (John 6:35), 48; 10:10; 11:25), and black people aspire to share in the fullness of that life. It would find it insulting to call on the black community to adopt attitudes of humility and self-abasement in the face of white oppression. The task, rather, is to do repentance for its sins of neglect in not challenging American racism. C. S. Eric Lincoln observes that "white theology is an entrapment that leaves the Black

---

87. Robin Scroggs, "Sociological Interpretation of the New Testament," in *The Bible of Liberation: Political and Social Hermeneutics*, ed., Norman K. Gottwald (Maryknoll, New York: Orbis Books), 1983.

Christian without hope" and results in white Christians deluding themselves about their Christian responsibilities.[88]

If black theology were to be classified in terms of a theological model, it would share the description of anthropological model with liberation theology, for it emphasizes that theologizing is authentic only when it is rooted in the history of the struggle of the poor and the marginalized, and that God and Christ continue to be present in the struggles of the people to achieve full humanity. There is, for black theology, an intimate and necessary relationship between eschatological salvation and historical liberation. The black community invites all, both black and white, to produce this liberation by their effort, cooperation and understanding. They offer the challenge of white repentance within the context of freeing the black community from its state of repression. William Jones says:

> White churchmen, if they would be faithful to the Gospel, must come to terms with the inextricable relationship of love, power, and justice...Reconciliation without justice is impossible...the white church must exercise repentance and reparation.[89]

Black Theology is Future-Oriented

Just as Jesus points to the future Christian environment which He intends His words will produce, so black theology looks to a renewed and, in some ways, radically new Christianity. Black theology actively awaits the day when racism in America will be perceived by whites as that which degrades and nullifies their Christianity, and will do the soul-searching necessary to remove its degrading influence. It is the future to which the Gospel message points. The black theologian, as the Marxist, is not satisfied to leave the future in the hands of those who now control the present, because the present remains unjust and racist. Of course, the call to change the present with an untried future meets with resistance. The resistance of a society satisfied with itself to the claims of black theology is typical of that of any society under scrutiny by the marginalized groups in that society. If black theology claims that the white church is not representative of Christianity, this can have a traumatic effect on those white Christians who have spent their entire lives devoted to the church of their

88. C. Eric Lincoln, *The Black Church Since Frazier* (New York: Schocken Books, 1975), p. 145.
89. William Jones, Jr., *God in the Ghetto* (Elgin, Illinois: Progressive Baptist Publishing House, 1979), p. 57.

choice. If they become convinced that the accusation has validity, then they become open to seeing no future prospects for their own religious development. Even those most open to discussion generally will tread cautiously before entering into a dialogue that could possibly shatter their vision of what church has meant to them since childhood. Under the influence of a future-oriented black church, the future of the American white church could take on a very different configuration. The earliest stages of an impact of the black church upon the thinking members of the white church would be marked by doubts and antagonism. The challenge that black theology offers to classical theology is clear and unambiguous. It says: "Take the Gospels seriously." The Black theologian believes that when this day comes, America will begin the task of purifying itself from racism.

The Question of Vicarious Suffering

The point of departure for black theology is the black experience, the experience of slavery, humiliation and suffering. While classical theology has occupied itself with such questions as the existence or non-existence of God, the Christological questions of the early Councils, or metaphysical explanations for the problem of evil, black theology has raised the question of liberation and survival in a hostile white Christian environment. Metaphysical speculation is not the first priority of a people examining the question of why it is that the followers of Jesus oppress and enslave. The black experience becomes a source for black theology when, upon reflection, the black theologian sees the divine will being worked-out in the lives of the black race. The relationship between their suffering and the conversion of the white race is never far from the consciousness of most black theologians:

> Slavery was but the means for inextricably binding the Negro and the Caucasian. Without this binding the immeasurably more bruising work of releasing whites from their blasphemous bondage to whiteness and racial superiority cannot be done.[90]

Washington contends that the doctrine of vicarious suffering is the only biblical model that can account for the black experience in America. For him, the black man is the contemporary suffering servant. Most black

90. Joseph Washington, *The Politics of God* (Boston: Beacon Press, 1969), p. 170.

theologians of the past agreed with him, but there are some notable exceptions. Cone contends that there is no definitive answer to the history of black suffering, but that there would be no need for a Christian doctrine of salvation if there were no evil in the world. This is not an explanation, but it does give us a theological trail to follow. In *The Politics of God*, Washington produced what may be called the first systematic black theological statement, in an academic sense. While black religion in America had always been a religion of protest, the emerging black theologians focused on "black" as determining the main feature of theology in the black community. Not that black theology is a treatise on the color of God, but the nature of God is revealed in His "color." The color of God only assumed such vital importance because color played a major part in the determination of human intelligence, human privilege and human value in American society. The question is not whether God is physically black, but whether a man who is black can identity with a white God and can count on His love and protection. If the suffering in the black community derived from color, then the theological explanation of that suffering had to have a strong color component. This is especially true in the black community because it treats religious questions in a very personal way. For God to be personalized in the black community, He must be "black." God must be familiar with, and vitally interested in, the oppression He experiences through His suffering children in the streets of the black community. Relief from suffering cannot possibly come form a "white" God. "White" is the symbol of His oppression! The black experience, therefore, is both a source and an essential component of black theology.

Why Black Theology?

The statement by the National Committee of Black Churchmen on June 13, 1969 was inspired by the black theological work which had been done to date; it inspired the black theological work which followed. To the questions about black theology, it responded:

> Why Black Theology?
>
> Black theology...emerged from the stark need of the fragmented black community to affirm itself as a part of the Kingdom of God...Black theology was already present in the Spirituals and slave songs and exhortations of slave preachers and their descendants.[91]

The etiology of black theology is difficult, not because its roots cannot be traced, but because in tracing those roots one is obliged to move away from the normal theological discourse and become acquainted with, and put into theological perspective, the history of the black race in America. The theological issues are connected with the socioeconomic issues which, in turn, are connected with theological issues. This is not new in the history of the development of theology. Many theological issues have resulted from a response to historical, social and economic events. The Papal Encyclicals and their theological commentaries are an example of how theology evolves out of issues that are important to a given people at a given time or place. The event and documents of Vatican II have encouraged a more anthropocentric thrust in theological writings. Black theology has evolved out of the need to put in some sort of theological perspective the history of slavery and the problem of how it was that a white Christian community engaged in it. There are related problems to which black theology is a response, together with other academic disciplines. The whole question of the value of cultures which are not the dominant cultures in a society has theological implications, and what black theology affirms is that in white theology God has been made identical with the culture of white Christianity.

Black Theology and Black Power

Black theology, as an academic discipline, began with the emergence of the Black Power advocate as he confronted American racism, inspiring a theological counterpart to the Black Power attack on the white political establishment. Black theology was and remains a polemic hurled at the American white church leaders and their theological traditions. The statement of the National Council of Churches Conference on the Church and Urban Tensions, which took place in Washington, D.C., September 27-30, 1967, made some references to the "Black Power" connection with witnessing to the Christian message. The Conference had decided to make the unprecedented move of dividing into two caucuses, one black and one white, each issuing different Declarations. The Black Caucus called on the white churches to admit that in spite of many good deeds in the past, the white churches, at that point in time, had not as yet affirmed the legiti-

---

91. *Statement by the National Committee of Black Churchmen*, issued at the Interdenominational Theological Center, Atlanta, Georgia, June 13, 1969, paragraph No. 3.

macy of the Black Power movement.

The White Caucus issued its own statement, acknowledging that the cause of the black theological community and the Black Power Movement was just, and that white America needed a change of heart. It acknowledged that the black ghettos were created not by the black man but by the white man, a fact that called for some sort of reparation.

The passionate language of black theology, and even its name, were not derived from moderates like Martin Luther King, Jr. but by the Black Power Movement, which represented the end of black-white consensus and coalition in the politics of the major United States cities. The spill-over effect was that black theology also moved away from the process of black-white consensus and coalition in the churches and theology in America. Black Power proclaimed the black value and cultural system that had so long lived under the thin veneer of the white cultural world, and black theology asserted the power and the vitality of black religious life, a life which survived over one hundred years hidden within the powerful stanzas of the black Spiritual. So much did white America associate the expressions of black theology with those of black power that, according to James Cone:

> Almost all wanted to know what was the basic norm in my perspective of black theology—black power or the biblical witness to Jesus Christ.[92]

Cone says that he refused to answer the question because the urgency of the time in which he lived demanded that his energies be placed elsewhere. It was in the cry of black power that Cone and others found the word of God. Not even in the relatively passive knowledge of black needs found in the black church did the black theologian find his God-language. The truth which resided in the black church needed to come alive with the prophetic fire found in the Bible; it needed to take flesh in the men and women of the black community, and it was the advocates of Black Power who best articulated this word. For Cone, God's word is found in the ghettoes with those who have no power to defend their humanity; their cry is the Old Testament prophecies come alive in our time. Cone credits the Black Power movement for clarifying in clear and prophetic language these black needs and aspirations.

---

92. Cone, *For My People*, p. 32.

Jesus is Black

Albert Cleage, whose strident language gave form to the black Christian nationalism, influenced by his association with Malcolm X and the black radical members of the Student Non-Violent Coordinating Committee, made a direct connection, insisting that "church" can only be predicated of those religious groups who became an integral part of the movement. The black clergy, in general, and the black theologians in particular, felt that Cleage was exaggerating and distorting, but that he was making a point, so some supported him and liberally quoted him from the pulpit. His theology was not acceptable to other black theologians because it reduced the Christian Gospel to a literal identification with the ideology of black power. Cleage thundered:

> When I say Jesus was black, that Jesus was the black Messiah, I'm not saying, "Wouldn't it be nice if Jesus was black?" or "Let's pretend that Jesus was black," or "It's necessary psychologically for us to believe that Jesus was black." I'm saying that Jesus WAS black., There never was a white Jesus.[93]

While keeping a "theological distance" from Cleage, the major black theologians honor and respect him. They understood the real meaning of his rhetorical language, his exaggerated examples, and they thought it odd that the white churches fear the strident and threatening voice, for such was the voice of the Old Testament prophets. Black power and black theology became wedded in an atmosphere of heady joy and passionate religious expression, and Black believers became imbued with its magnetism. For the black theologian, the dethronement of all the evil powers which had oppressed the black man since slavery and the victory represented by the resurrection is the eschatological destiny of blackness. It identifies with blackness because an affirmation about this particular people and slavery is related to the mystery of the cross. The mystery of the cross, for the black community, is not a seminary subject, but a reality which is experienced in the street and in the workplace. Blackness is skin color and blackness is capsulized history of a people. This duality is affirmed by black theology and interpreted theologically, homiletically and liturgically in the life and worship of the black church. Black Power fed, and was fed by black theology. The anger of the streets was refined, absorbed,

---

93. Quoted in Alex Poinsett, "The Quest for a Black Christ," Ebony, March 1969, p. 174.

and returned to the people in strong but academic theological language. Some white churchmen applauded the twinning of black power and black theology. Kyle Haselden wrote:

> We must ask ourselves whether our morality is itself immoral, whether our codes of righteousness are, when applied to the Negro, a violation and distortion of the Christian ethic...[94]

Black Theology and Liberation Theology

It is clear from the demands of black activists and the statements of black theologians that black theology in the United States has been influenced by Latin American Liberation Theology. The relationship between the two has not been adversely affected, but mutually challenged by the fact that liberation theology takes class as its point of departure and black theology takes color as its point of departure. Like liberation theology, black theology is committed to the dual liberation from sin and guilt and from physical oppression in all its forms in American society. In both disciplines, there has been an attraction to Marxism, whether for critical analysis of the capitalist system or for the vision of a new society stripped of the evils that have accompanied the development of Capitalism. Black people, like the campesinos of Latin America, discover their Christianity from the bottom rather than from the top of the socioeconomic ladder, from the revolutionary movements that fight for justice, rather than from those whose task it is to preserve the theological status quo. What is "holy" in the religion of the powerful and comfortable is most often the celebration of the sacraments and the Eucharist. But, "holy" for the oppressed peoples can take the shape of a radical challenge to the injustices inherent in particular social structures of power. The black and liberation theologians interpret the Gospel message as the refusal to accept the things that are as the things that must be. This refusal is part of the content of black theology. It is an affirmation of Jesus as Liberator of the world and all its peoples. Both disciplines share the view that, since Constantine, the Christian hierarchy have mistakenly advocated a spiritual view of the Gospel that separated the confession of faith from the practice of political justice. Cone emphatically challenges the assumed separation between "faith and political praxis," pointing out that liberation theology in all its forms (black

---

94. Kyle Haselden, *The Racial Problem in Christian Perspective* (New York: Harper and Row, 1959), p. 48.

theology, feminist theology, African theology) rejects the classicists' dichotomy, and insists that there is between faith and political praxis a dialectical relationship.

Black theology and liberation theology both refuse to characterize Christian theology as abstract, dispassionate discourses on the nature of God in relation to man. For black theology, theological language is passionate language, the language of commitment, for it seeks to vindicate the afflicted and condemn the enforcers of evil. Black theology joined the liberationists of South America in their attack on Capitalism, since the black church has been preoccupied with its own survival, but it did not take the lead in analysis. The need for institutional survival left little time or inclination to attack what James Cone now sees as an important obstacle to black freedom, the lack of an economic agenda. The black church, therefore, now seeks to study other socioeconomic models in which the black community might develop fully. One such model was offered by a Baptist preacher, the Rev. George Washington Woodby, pastor of the Mt. Zion Baptist Church in San Diego, California. In the early 1900's he was a prominent Socialist and his book, *What to Do and How to Do It or Socialism vrs Capitalism* was widely read and commented on. Woodby made his attempt in the context of the Socialist Party. Cornel West offered a different approach. He suggests using principles espoused by the Marxist, Antonio Gramsci:

> Based on the insights of Gramsci....I shall present a theoretical framework that may be quite serviceable to black theologians, Latin American liberation theologians, and Marxist thinkers.[95]

Cornel West and the Counter-Culture

West suggests that the black community offer a counter-culture to America with the hope that it will influence the rest of America, just as the civil rights movement initiated by blacks influenced a wider community in raising consciousness on civil rights matters. Just as whites now sing "We Shall Overcome," so a black movement of a counter-culture could spread to the rest of America. Its legitimation must take place in the cultural and religious spheres if it is to maintain itself over a long period of time. West chooses Gramsci's writings as a model because it proposes that no state

95. Cornel West, "Black Theology and Marxist Thought," in *Black Theology: A Documentary History*, p. 563.

and no society can be sustained by force alone. Gramsci replaces Marx's ideology with the concept of hegemony, a process of ideas, beliefs and behavior that becomes the status quo. Black theology wants to go beyond observing the moral implications of socioeconomic practice to proposing its own answer to such practice. In this, black theologians feel they are answering a need not responded to by classical theology. True to its long tradition going back to Africa, black religion has never accepted that secular matters are someone else's business; black societies have never seen wisdom in separating the secular from the sacred. In returning to its African roots, black theology has redisovered this aspect of African religious life, and is in the process of adapting it to the black diaspora, wherever it may reside. The black theologian recognizes that the truth in the Gospel of Jesus is a truth that must be done and not simply spoken or preached. Of preaching we have seen and heard enough. The act of denunciation of injustice can be a "word" given effortlessly and from a position of security. What the Gospel calls for is a life of action to bring life to the words we speak, and the sermons we preach. It is precisely here that, for the black theologian, the white American theologian has failed. The tradition of separating the secular from the sacred is so imbedded in Classical theology that Cone and West see little hope that such a theology will make the transition from contemplation to action. It is for this reason that West feels the black community can make a significant contribution to restructuring those areas of American society that still produce unjust structures and practice. Classical theology, inhibited as it is in entering the political and economic areas, will not be able to do this. Black theology has taken up the challenge as its own unfinished agenda. Ironically, it is not as resource-equipped to take up this enormous task as is the white theological community, but impossible tasks are part of life in the black community.

Lee Cormie, of St. Michael's College of the University of Toronto, and coordinator of the Theologians Project of Theology in the Americas, makes the connection between black theology and the critique of Capitalism:

> ...in the late 1960's liberation, as distinct from equality, emerged as a central theme in black theology...In particular, there was a call to reject dominant forms of theology, and to articulate a new theology expressive of the experience and faith of oppressed blacks...this kind of liberation theology must be understood ultimately as a challenge to the capitalist system in the U.S.[96]

Black theology is a critique of classical theology, which it sees as a theology buttressing an economic system that marginalizes and degrades members of the black community. Black theology is a political theology because it is the practice of politics that gives some hope for changing the structures of unjust systems. Faith, religion, piety, theology—all are intimately connected to, and involved in, political change. But its approach to changing structures is quite different from the Marxist approach which it has sometimes used. Black theologians recognize that cultural and religious attitudes have a life of their own not easily capsulized in terms of class analysis.

Black Theology is Survival Theology

Black theology concerns itself with the survival of the black race, so it is survival theology. It provides the theological dimensions of the dialogue that black people have with each other when they meet to discuss survival in a land that denies them equality. Together with the institutional black church it provides a "meeting place" where black people can discuss the means by which they can fully emancipate themselves. As survival theology, black theology proclaims that anything that supports or comforts or rationalizes white oppression is anti-Christian. This is true even if that support or rationalization emanates from the precincts of a church. As survival theology it proclaims that it seeks to find a God who identifies with the need of black people to survive in an atmosphere of oppression. It searches for a new language in theology. "God is black," "God lives in the ghetto." As in any course in the techniques of survival, the role that black theology has assigned itself is to instruct black people in that which is necessary for the survival of both body and soul.

The black church has always been the teacher and protector, the comforter and the enlightener of the black community in America, and those most articulate in the black church (the black theologians) see themselves as obliged to be an extension of the survival teachers in that church. James Cone, in a statement that proved to be very controversial, had this to say:

> The role of black theology is to tell black people to focus on their own self-determination as a community by preparing to do anything which the community believes to be necessary for its existence.[97]

---

96. Lee Cormie, "Liberation and Salvation: A First World view," in *The Challenge of Liberation Theology*, p. 33.

The National Conference of the Black Theology Project, 1977, strongly influenced by the presence of James Cone, stated that the main issue for the black race (and therefore for the black church) is survival. The conference committed itself to the struggle for freedom from injustice and racism. In the process of teaching the Gospel values to their people, the black theologians and the black church are very careful not to do so in such a way as to blunt the impulse for black freedom and full humanity. The black church is careful to balance its eschatological promise with the present need for survival. Since the black church is the only institution over which black people have total control, it sees itself as dutybound to continue its leadership role in the struggle for full black participation in American life.

Black Theology and the Theology of Hope

In the late 1960s many black theologians studied the writings of Jurgen Moltmann, as a representative of the Theology of Hope, and found points of agreement. Both shared the basic assumption that God's Revelation continues to express itself in present as well as past history. Both reject a theology which gives religious sanction to the values of a nation. But, a major criticism of Moltmann's theology is that it is too futuristic. DeOtis Roberts explains:

> ...his view of revelation is too narrow. Eschatology...(must) be realizable eschatology in the here and now in order to be hopeful to black Christians.[98]

And yet, the black theologians appreciated that Moltmann was quick to recognize the contributions of black theology. They observed that he spoke out forcefully against racism and did not hesitate to say that black people and other oppressed peoples have the right to pursue their goals with more aggressive means when their nonviolent efforts to effect change have proved fruitless. Moltmann reinforced the black insistence that God does not expect the black man to accept the present reality of things. This could not be His intention for Humanity. Theology can only be understood as insisting that the present condition is incongruous with

---

97. Cone, *Liberation*, p. 41.
98. J. DeOtis Roberts, "Black Consciousness in Theological Perspective," in *Quest for a Black Theology*, ed., James J. Gardiner and J. DeOtis Roberts (Philadelphia: Pilgrim Press, 1971), p. 81.

the expected future. To Moltmann's credit, he understood that a rejection of elements of his theology had as much to do with the need for the back man to develop an indigenous theology, as it had to do with any fundamental disagreement on the question of the future. Moltmann's sensitivity to the need for the black community to speak for itself was apparent at Duke University on the night of Martin Luther King, Jr.'s assassination, when he declined to answer the question as to how his European theology might respond to that tragic event. Although he avoided direct criticism of black theology, it is possible to see in his writings points of disagreement. He maintains that human emancipation cannot be actualized without a penetrating analysis of economic and political reality followed-up by an active praxis. Moltmann's analysis is right to the point and explains why black theology has been so slow to dialogue with other liberation movements and why, privately, it has had such difficulty in coming to an understanding with the liberation theologians of Latin America. Cone criticizes the Theology of Hope for not giving sufficient attention to the "present." The criticism may be misplaced since Moltmann's concern is to avoid the deification of man. DeOtis Roberts was less critical agreeing that an eschatology without a future dimension is not complete. Moltmann questions whether a theology which is militant can effectively produce reconciliation, which he considers necessary for a humane society. He decries the condition of a society in which men differentiate themselves on the basis of race, but at the same time he insists that theology must offer the hope of reconciliation and that means it must take steps to bring it about. He approved of Martin Luther King's decision not to use the methods chosen by his opponents and not to seek power and its prerogatives. The revolutionary "atmosphere" causes Moltmann to be concerned that violence for its own sake may be adopted by the revolutionaries. He writes:

> Radical Christianity will have a revolutionary *effect*, but a revolutionary program would be just the way to neutralize it. The title "Revolutionary" must, if at all, be given from the outside; one cannot claim it for himself.[99]

While it was Moltmann himself who contributed to energizing the thinking of black theologians like Cone during their formative years, it is clear that there remains a wide gap between them. And unless the ques-

---

99. Jurgen Moltmann, *Religion, Revolution, and the Future* (New York: Scribner's, 1969), p. 40.

tion of rhetoric and sweeping generalities is somehow settled, the chances for serious dialogue between the black theologian and any of his white counterparts remain problematic. However, in spite of these problems, there are individual theologians who reached out to understand. The theologian, John Carey, writing in *Theological Studies*, found the value of black theology in its having made quite clear the important role of the black experience in the American Christian tradition.[100] This kind of recognition plus a willingness to understand the historic realities that have produced black anger will go a long way to effecting reconciliation. At the same time, the black theologian must put a limit to the extent to which his anger affects his judgements.

100. John Carey, "Black Theology" in *Theological Studies*, September, 1974, Vol. 35, p. 520.

CHAPTER FOUR

MAJOR BLACK THEOLOGIANS

The Major Theologians

We have discussed the sources and content of black theology without particular reference to the major black theologians themselves. To flesh-out the subject of black theology and to contrast the differing approaches to the subject among black theologians, this Chapter will focus on four major black theologians: Joseph Washington, James Cone, J. DeOtis Roberts and William R. Jones. Each of these men represents one of the four major streams of black theology. Joseph Washington represents an early black theology that made every effort not to stray too far from mainstream classical theology. James Cone represents black theology's attempt to characterize classical theology as unchristian and in need of radical transformation. J. DeOtis Roberts reflects on the need of reconciliation with classical theology in the struggle for black liberation. William R. Jones accuses all the other major black theologians of having assumed the validity of the theodicy of white classical theology, and that this fact negated the so-called radical change in theological method supposedly initiated by black theology. The black theologian's assumption that God is on the side of the oppressed, that He loves black people is questioned by Jones who asks the question: "Is God a White Racist?"

Each of these men has made a major contribution to black religion and the literature of black theology and, in the process, has come closer to the construction of a genuine American theology than has the enormous proliferation of theological faculties in white America. They struggled to remain "Christian" in an atmosphere in which they reached down deeply into their passions to excoriate the failure of white Christianity to bring about a condition of brotherhood with the black community. They hope to bring about, in the future, a society in which children will begin their lives free from the burdens of racial misunderstandings and antagonisms. Each of them, in his own way, addresses himself to the "why" of black theology. Alan Boesak reflects their reasoning:

> The "why" of black theology is not difficult to understand, Until now, white Western Christian theology lived under the illusion that it was a *universal* theology, speaking of all those who called themselves Christian. Christian theology has been cast in a white Western mold...It did not reflect the cries and the faith of the non-white poor and oppressed.[101]

Joseph Washington

Commentators on black theology usually begin with the works of Joseph R. Washington, Jr., especially *Black Religion*, published in 1964, and making the point that Black Church is a fraternity of sociopolitical actions and not a church with a theology, and *Politics of God*, published in 1967, in which the author stresses the need for combining political action with religious practice. Prior to the publication of these works black religion was considered as an extension of American Protestantism. Washington devoted much of his time, writing, and energy pointing out that black religion was a social and religious style of worship that differed in many and significant ways from the religion of the white denominations. He went so far as to say that black religion was so preoccupied with the hope of freedom from oppression that it failed to develop a proper black theology. Washington went even further by claiming that, in its preoccupation with strategies of freedom, the black church could scarcely be called a Christian community, since it uses the church as a base for political ends. Its sophisticated theology, he says, is focused solely on the destruction of the black ghetto.

The suggestion that black religion had deteriorated at the expense of forays into the areas of civil rights raised a storm of controversy amongst black intellectuals and preachers. Washington was strongly and emotionally challenged by most of his fellow black theologians. They were enraged over his claim that the black church was hardly church and certainly had no theology. Cone commented:

> Although *Black Religion* was received with enthusiasm in the white church community, receiving major attention from white scholars, it was strongly denounced in the black church community...Indeed, black theology, in part, was created in order to refute Washington's book.[102]

---

101. Boesak, p. 55.
102. Cone, *For My People*, p. 9.

Reflection and the passage of time has softened the criticism of Washington by his fellow black theologians as they revise their own perceptions of the temper of his time and the prevailing theological models. They do applaud him for criticizing the white churches for excluding blacks from the mainstream of American Protestantism. But they fault him for maintaining that, having been excluded from mainstream white Christianity, blacks were without theological content in their religious expression, or that their churches were not genuine manifestations of Christian religious life. Among his many statements that angered black theologians, Washington's contention that a church without a theology is a contradiction in terms and his application of that statement to the black church, caused a furor. Washington appeared naively unaware and perhaps ignorant of the fact that the social protest found in black religion from the time of slavery was a genuine historical manifestation of Christian praxis. He was, in effect, negating the entire history of the interplay of religious and liberation forces and sentiments in the history of the black church. And even more, Washington had not properly understood the African tradition (which the slaves carried into their religious meetings) of not separating the secular from the sacred in their practice of religion. Cone believes that it would be a serious mistake for the black theological community to depart from the continuity of the secular and sacred so prominent in African social life.

Oddly enough, Washington's claim that black religion was not really a manifestation of Christian religious expression served as stimulus in calling forth from the black intellectual community the desire to delve into the question of the relationship between religion and social activism. Cone remembers:

> If we were going to protect the black community from Washington's promotion of this widely held white thesis that the Black Church has no theology, then we needed to give some intellectual structure to the implicit Black Theology that we claimed was already present in the history of the Black Church.[103]

The Suffering Servant

Washington advanced the theory of the "suffering servant," i.e. the black man's role in converting the white race, in bringing about its justification

---

103. Cone, in *Black Theology, A Documentary History*, p. 610.

by his history of suffering. If there was anything genuinely "religious" about life in the black community, says Washington, it was the fact of the acceptance of this role in the Divine scheme of things. However, the black man should never again aspire to participate in matters religious with whites. The "suffering servant," he wrote, must not participate in the socio-cultural life of the white man. Washington did not deny that some whites also suffer for other members of the Church worldwide, nor that their suffering was inefficacious for redeeming others, especially their own white brothers and sisters, but he considers their suffering a luxury which they can afford to choose if they so wish. He says:

> He (the Black Minister) can know that in this singular issue whites who join with Negroes in this warfare are sufferers as well, but their suffering is one chosen by them which can be renounced, while the suffering of the Negro is not of his own choosing and cannot be escaped.[104]

Washington's attempt to explain black suffering in such a way as to satisfy the black community without alienating the white community led him into theological positions which alienated him from the back theologians and opened him up the charge that his understanding of ecclesiology was deficient and his grasp of the Gospel message itself lacking in depth. His position was somewhat strengthened by the fact that the theory of the "suffering servant" was accepted by Martin Luther King, Jr., but King himself was criticized on this matter by, among others, William R. Jones and James Cone. Jones maintained that the criticism of Washington was also applicable to any black theologian, including Martin Luther King.

King did call for the radical reconstruction of society, a position Washington would have had some difficulty in adopting, but this did not prevent him from coming under attack for refusing to take militant steps in the pursuit of freedom. Washington did not wish to contemplate a theology without the "suffering servant" theory. The alternatives were too painful. Could white insensitivity to black men mean that Christianity has been seriously flawed from the beginning? Could God be, as William R. Jones was to ask, a white racist? The only explanation for the history of black suffering that would satisfy Washington was that blacks had been chosen as God's children, His most favored folk. He resented any white pretensions in sharing the title of "chosen":

---

104. Washington, *Politics*, p. 186.

> It is true that Christians claim themselves to be God's "new Israel," called to be his "suffering servant" people. But in the Western world of whites this claim has been reduced to a mythical hope with precious little relationship to reality...Whites can hardly claim to be "stricken, smitten" because they are whites or Christians. Negroes can only perceive themselves to be afflicted by God.[105]

Washington concedes to a less than solid theological validity to his theory of the "suffering servant" by agreeing that this understanding of the role of the black man is based on an interpretation of the historical experience of the black man rather than on some tested and convincing religious or theological insight. This concession comes in his observation that even the Israelite conviction that the Jews are the chosen race comes from such a method. He says;

> That the Israelites were the chosen ones is a declaration of faith made by the interpreters who in historical perspective have asked and found the meaning of the paradoxical and enigmatic life of the Jewish people.[106]

Washington A Theologian of Balance

James Fowler, Professor of Theology at Emory University, listed Washington as one of the "Theologians of Balance." These theologians are contrasted with the "Ideological" theologians who see things in terms of dichotomies, as for example, reason versus revelation, oppression versus oppressed and white versus black. He defines the faith of the "Balance" theologians as conjunctive, joining together what has been separated, viewing as complimentary what has been previously been seen as opposition. The oppressor and the oppressed are not seen as antagonistic but rather as people who both experience adversity, sometimes among themselves and sometimes with others. In the same manner, black and white compliment each other presenting truth in a wider dimension. The passion of the "Balanced" theologians is infused with reason. Fowler describes them as dedicated to discover God's work and will in history, but cool and objective in their interpretation of their intellectual discoveries. He includes Washington among their ranks. For those black theologians who see religion and politics as necessarily intertwined, Washington's claim that the black church has no theology and has not produced any theolo-

105. Washington, *Politics*, p. 155.
106. Washington, p. 156.

gians comparable to Paul Tillich or Karl Rhaner aids and abets white theologians' claim that the Gospel teaching should be separated from revolutionary struggle. Such an acceptance and ratification of the viewpoint of classical theology denies the validity of the history of the black church as a church of protest against the prevailing socioeconomic condition of the black community.

Washington's writings made it imperative that his thesis be challenged by other black intellectuals and men of religion. Black theology, then, by having the need to respond to Washington, reinvigorated itself. Washington did not claim that blacks had no religion, but rather that the religion they practiced was not the "Christian" religion as known and practiced by the white community. He drew the conclusion that black churches were "Christian" only by association, and were thus unable to present their religion as representative of Christianity. The church developed by the black man, he asserted, under the conditions of racism and slavery, did not engage in a theological dialogue that represented that of the ancient Christian tradition. It used the "Christian" form to present its demands for social and economic equality. It was a strategy. The black clergy contended that Washington was confused. It was black religion and black religion alone, in America, that was truly Christian. It was black theology and black theology alone that was a truly Christian theology, since it alone exposed the sins of racism and slavery. It was black religion alone, they claimed, that had integrity, since whites preached brotherly love but did not practice it, while blacks practiced it to a very high degree in their readiness to forgive the white man and in their willingness to enter into theological dialogue and religious worship. Black theologians are quick to point out that although Washington was an early expounder of the uniqueness of black religion, he did not represent the rhetoric and the essential meaning of black theology. However they do not minimize his importance to the development of black theology, even if part of that importance was due to the negative function of motivating them to respond with a positive black theology. Moreover, Washington's demonstration of the uniqueness of black religion, its total independence from any of the white denominations, helped set the stage and the atmosphere for speaking about and developing a black theology. Also, they agreed with Washington that the black church in the ante-bellum period was more authentic than that which followed the Civil War, since it was committed to the Underground Railroad, slave insurrections and other activities which were considered illegal from the white perspective. That Washington's importance was constantly reviewed is reflected in Carlton L. Lee's comment:

> There is a strong suggestion that (Washington) has not thought carefully about many of the facts of the Negro-American community, nor does he interpret the facts with the understanding or depth that the subject deserves....It is unfortunate that the author...was unable to do what he most certainly wanted to do: say something relevant and significant about religion in the Negro-American community.[107]

The Assimilationist Model

Washington can be said to have begun his theological investigation into the nature of the black church by having proposed the assimilation-isolation model, a process by which persons of different racial and ethnic backgrounds come together to join in a larger community. The ideal in this model would see the entire disappearance of separate social structures based on race or ethnicity. The reconsideration of his position in his later writings indicates that the assimilationist model is irrelevant to a black church which desperately needs to seek its own identity and recover its own culture before it will be in a position to begin to assimilate, on a level of equality, with the white church. It could be that when Washington spoke of racial assimilation, he was really referring to acculturation, the cultural change that comes about as a result of frequent contact between two or more cultural systems. Acculturation differs from racial assimilation in that the latter is characterized by the ability of entering into primary relationships with the dominant group in society. The experienced missionary can testify to the fact that the formation of primary relationships between missionary and local people is very rare, if not impossible.

Washington's hopes for racial assimilation in America were premature. To expect racial assimilation in the midst of racism and the most serious social tensions was naive. More realistic is the opinion of those black theologians who believe that whatever kind of assimilation blacks and whites will be able to achieve, must be achieved *after* the black man has attained equal rights in the land of his birth. G. Clarke Chapman, a professor of theology at Moravian College, in Bethlehem, Pennsylvania, quotes Gayraud Wilmore in an interview he (Chapman) had with him.

---

107. Carlton L. Lee, a review in *The Christian Scholar*, Fall, 1965, pp. 242-47.

> Reparation means...to repair the damage I've done...in order for reconciliation to happen. Reparations, reconciliation, and forgiveness belong together, and the white church will have to face up to it...[108]

Washington on the Black Preacher

Joseph Washington examined the folk religion of the southern migrants and concluded that this religion, pregnant with the yearnings for social justice and characterized by protest, was betrayed by the moralistic and dictatorial preacher. The irrelevance of the black church, for him, was due to the calming of black protest by the preacher. He described how in that era of decline in the quest for freedom, the Negro minister, as spokesman for the people, preached moralities instead of freedom and emphasized rewards, while succumbing to the coaxing and bribery of the white power structure. Washington analyzed what happened when black folk religion was led by uneducated black preachers who sought favor with the white majority. He claims that, under these conditions, black religion lost its outward manifestation of protest. Cone, and others, against Washington, claim that in spite of the conditions described by Washington, blacks retained religious sophistication and continued to manifest their anger and protest in many and sometimes ingenious ways. Not all observers of the history of black theology are as hard on Washington as is James Cone. Sontag and Roth see Washington as in the mainstream of the black power tradition, contrasting him with a more moderate King. Their characterization of Washington does not appear exaggerated when one considers Washington's own rhetoric:

> ...the task of Black Power is radical change in the society to "smash racism." Outside the revolution there is no Black Power.[109]

In the early sixties, Washington continued to hold onto the hope that a real transformation in race relations might still be possible. By the late sixties, however, he was less optimistic. He later believed that only black religion has the hope of saving authentic Christianity in America. Washington's move from hope of reconciliation to its abandonment brought him into the mainstream of contemporary black theological thinking.

---

108. Quoted in G. Clarke Chapman's "Black Theology and Theology of Hope," in *Black Theology, A Documentary History*, p. 198.
109. Washington, *Black Power and White Subreption* (Boston: Beacon Press, 1969), p. 118.

James Cone

"Black Power" was one of the most provocative and revolutionary phrases of the 1960s. Young black Americans paraded openly with guns and threatened to use them against white America if their demand for equality before the law was not honored in fact as well as theory. Many in the black community identified with the rhetoric and concepts of black power. Self-determination, self-identity, self-help and the declaration that black was indeed beautiful became important elements of black thought and discussion. It was in the 1960s that black community leaders abandoned the hope that integration into the mainstream of American life could be accomplished without any real political or economic power. The black theologian who had regarded himself as belonging to the black elite, the vanguard of black thought and theology, now felt himself replaced by an articulate and militant group of young, angry blacks who neither claimed an educational background nor expertise of any kind. In a remarkable self-analysis the black theologian judged that his safe haven within the ivy walls of white theological institutions had caused him to get out of touch with what the black community was thinking and feeling, so he "took to the streets;" the black theologian became involved. It was in this atmosphere that James Cone emerged as a black spokesman in theology. He had barely received his Ph.D. in systematic theology at Northwestern University when he started to work on his *Black Theology and Black Power.* This was in 1965 and he immediately became the most prominent exponent of black theology in the nation and in the world.

Cone took an affirmative stand on the question of black power. He saw it as a challenge to the racism of white America and the complacency of the black church and its representative theologians, a prophetic voice in American society. For Cone, Black Power was an expression of Jesus's identification with the suffering poor, and unless it identifies with it the black church "will become exactly what Christ is not." Cone defines black power in such a way that the theme of revolutionary violence is openly advocated, an attitude which aims at emancipation by whatever means black people deem necessary. The importance of Christian theology, for Cone, is that it proclaims God's work in the world from the perspective of the Gospel of liberation as given to us through Jesus Christ. A community that fails to produce a theology of the oppressed is not a Christian community, for liberation is the essential message of the Gospel. The fact that Christ is the Liberator of the oppressed is the key to understanding the black struggle, to share in the benefits of being an American. Cone sees the task of black theology as that of bringing the message of liberation to the black community and stressing that it is "religious" to fight for its rights and that in that fight the God of the oppressed, who is actively involved in history, will be on their side. In an approving linkage between the Gospel and revolutionary activity, he says:

> That is why Camilo Torres was right when he described revolutionary action as a 'Christian, a priestly struggle.'[110]

The resurrection of Christ made freedom a sacred right, one that the Christian must risk all to attain, and theology must be completely immersed in the holy struggle. The language of theology must reflect its preoccupation with freeing the oppressed. It is a passionate project, pregnant with passionate language and prophetic cries. Theology cannot be neutral. Cone wisely pointed to white theologians in fortifying his argument that his presentation of black theology was based on essentially a biblical interpretation of the Christian faith. He called upon Karl Barth and Jurgen Moltmann, Paul Tillich, Albert Camus and Franz Fannon to show that a radical but historically accurate interpretation of the biblical story led to the inescapable conclusion that the affirmation of blackness against the dehumanization of white racism was a Christian imperative and that there could be no Christianity in America without it. Cone insists that the power of the black person to say *No* to white power flows directly from the message of Christ. Cone denied any disfunction between theology and black power. Rather, black power was basic to his interpretation of Christianity. He worked mightily to substantiate the social activism in the black community in a theological context. His work helped save black college youth from apostasy since they were, in his theology, the answer to their complaint that the black church was not significantly involved in the liberation struggle. Cone put the black church on the cutting edge of this struggle. Since Cone used blackness as the point of departure in constructing his theology, the question arose as to the relevance of the white person in his scheme of the Christian world. Cone perceptively asserts that being "black" in America has little to do with skin color. At this point it is a term used to identify solidarity with the poor and oppressed. Being "black" means that one's theological focus is in the ghettos of America.

---

110. Cone, A Black Theology of Liberation, pp. 19-20.

Cone argues that if the historical record is clear that God has always aligned Himself with the oppressed, then it follows that in order for a theology to be genuine and Christian it cannot do less. Thus black theology is genuine and authentic theology because it interprets the Gospel of Christ in light of the existential situation of blacks and proclaims clearly that the liberation of the black person in America is a high priority with God Almighty.

Cone's Response to White and Black Fears

To those who express the fear that Cone is substituting "liberation" and "black power" for Christ as the norm for black theology, Cone responds that it is important not to confuse the sources of black theology with the norm. Those sources are the black experience, black history, black culture, revelation, Scripture and tradition. To those who question his placing of revelation after black history and black experience, Cone responds that the numerical order in which he has placed his sources is not to be interpreted as the order of importance. He sees no particular order of importance in any case since they are all interdependent. Such fears are just another example, for him, of the bind into which classical theology has gotten itself, worrying about such priorities while omitting within its theological program some of the most important issues of our time. Cone defines Revelation as God making himself known to man through the historical acts of human liberation, as Yahweh's involvement in the Exodus. The weakness of white American theology, he says, is its inability to move past the first century in its analysis of revelation. Nor does black theology neglect Scripture in the theological discourse. In fact, it is a kerygmatic theology that identifies God as the God of liberation who promises that His righteousness will vindicate the suffering of the oppressed. Scripture, then is very important to black theology. But the white man's use of Scripture is not properly focused:

> While churches are debating whether the whale swallowed Jonah, the state is enacting laws of inhumanity against the oppressed... whites who insist on verbal infallibility are often the most violent racists.[111]

To those who express the fear that Cone is substituting "liberation" for Jesus, Cone responds that it is important not to confuse Jesus' behav-

---

111. Cone, *A Black Theology of Liberation*, p. 64.

ior in the first century with His presence among men in the ghettos of black America. Jesus, he says, is doing the contemporary work of God in the world, and is not limited by the limited visions of men who presume to interpret His will. Cone cautions that without the continuity between the historical Jesus and the kerygmatic Christ, the gospel becomes little more than personal reflections on the early Christian community. Cone is a militant theologian who does not accept the concept of a God who accepts the suffering of black people. This suffering, he says, must be taking place over His divine objections. The strength of his faith is shown here since his anger has not allowed him the luxury of ascribing to God, as William R. Jones suggests, the possibility that God does not care, or even worse that He is just another white racist. However, Cone does appear to be inconsistent here and to engage in a circularity of thought. After all, if he is unwilling to accept the concept of God as given in classical theology, then why does he explain black suffering by saying that God does not willingly accept it. The God who does not willingly accept black suffering is the same God as described in classical theology. It is precisely here that William R. Jones sharpens his attack on Cone. He does not allow Cone to go unchallenged when, on the one hand, he cannot conceive of a Christian perspective that does not stress liberation and, on the other hand, permits God to delay justice until the afterlife. Cone acknowledges the cogency of Jones' analysis. For Cone, it is after death that things are finally set right. The resurrection of Jesus Christ points to this reality and keeps it alive, enabling black people to give all for liberty now. Their reward, he says, is eternal. It is not surprising that Cone's eschatology is not accepted as internally consistent with the rest of his writings, since he also maintains that the idea of heaven is irrelevant for Black theology, and advises Christians against wasting their time contemplating the next world, if, in fact, such a world exists.

J. DeOtis Roberts sees Cone as more consistent with the teachings of the Black Muslims. In commenting on the denial of the afterlife by the Black Muslims, he makes this reference to Cone:

> It is ironic that James Cone, as Christian theologian, should be in the same camp regarding eschatology.[112]

---

112. Roberts, *Liberation and Reconciliation*, p. 162.

Roberts felt that Cone was over-rated as a black theologian, that he was not sufficiently open to learning from black sources and that his theological works were being produced so rapidly as to preclude depth of understanding. J. H. Jackson, President of the National Baptist Convention, which has a membership of over six million people, was not happy with a growing tendency on the part of Cone to suggest that more time be spent by black theology in the study of Marxist analysis. He saw no need for black theologians to go to Europe to borrow ideas from Karl Marx, a man who did not live the black experience. For Jackson, the use of Karl Marx as a tool of black theology turns black theology into a theology of polarization. And he objects to the symbol "blackness" saying that it too adds to the atmosphere of polarization. Jackson tied-in Cone's use of Marx and the use of the term "blackness":

> When he arrives at the color of Jesus, Professor Cone does it by the hands of dialectical materialism.[113]

Cone's style sometimes leads to misinterpretation, sometimes to inconsistency. He drew heavily upon humanistic existentialists like Camus. His doctoral research centered upon Karl Barth's doctrine of man. He read Marx and found much there that he felt could be incorporated into the black theology of liberation. He was influenced by the rebellious and defiant spirit of Camus which appeared to present even the most insurmountable odds. His training was along the line of classical theology. Thus, there are reasons why his theological program does not always run in a straight line. Perhaps *that* would be inconsistent. His parallels were with the slave condition of the Israelites in Egypt and with the oppression of first-century Jews and Christians under the Romans. It was in this context that he developed a theology that was a strategy for freeing the black body, but also the black mind from the beliefs and attitudes which blocked the movement toward liberation. The cause of liberation took priority over any "suffering servant" claims of the gospel or of other black theologians. Cone is slow to develop a social theory to compliment his incisive criticism of the role that white Christianity has played in the pitiful condition of the black man. Here again, the focus he places on the afterlife in the final solution of the black man's problems, leaves him open to the charge

---

113. J. H. Jackson, *Nairobi: A Joke, a Junket, or a Journey* (Nashville: Townsend Press, 1976), p. 75.

that he "talks about" oppression but does not "do" anything about it. His major weakness therefore, as I see it, is his lack of a social theory. One can forgive and understand the militancy, given what we know of black history in America, but it is difficult to accept that militancy when, in the end, he appeals to a nebulous eschatology as the answer to the black man's condition. For Cone's theology to be rounded-out it needs a social theory that will help resolve the difficulties inherent in the structures of oppression. Whatever the criticism of Cone, however, he remains the chief spokesman for black theology, as his voice remains incisively prophetic. To the question, Is Justification by faith enough? he has answered strongly in the negative.

J. DeOtis Roberts

J. DeOtis Roberts became a prominent figure in black theology in the 1970's, with the publication of several texts and many articles, the most important of which are, *Liberation and Reconciliation: A Black Theology* (1971) and *Black Political Theology* (1974). He is best known for his emphasis on reconciliation along with liberation in contrast to, for example, James Cone, whose focus is fixed on liberation. For Roberts, a theological project is authentic only if it allows for the co-equal ranking of liberation and reconciliation as theological goals. For fear that the black community could collapse into a condition of despair and quietism, he insists that:

> Against the consciousness of unjust treatment, an unloving relationship based on racism...there is a need to believe that God is just, loving and merciful.[114]

Roberts was strongly influenced by Jurgen Moltmann's *Theology of Hope* and Metz's *Theology of the World*. He found in these German theologians the theological foundation for the liberation theme which dominates black theology. They provided "respectable" support for the black theologians' insistence that God is in solidarity with and favors the poor. References to German theologians on the part of black theologians gave the black theologians the opportunity to get the attention of white American theologians, since European theology has been the source and inspira-

114. J. DeOtis Roberts, "Black Consciousness in Theological Perspective" in Quest for a Black Theology, ed., Gardiner and Roberts (Phila.: The Pilgrim Press, 1971), p. 71.

tion of the white American theological community. The hope was that those European theologians who were accepted by mainstream classical theology would inspire the North American theologians to assist black theology in exposing and eradicating the evil of racism in America; but that did not happen. Moltmann alluded to this at a New York conference on *Hope and the Future of Man* (October 8-10, 1971):

> Whose future do we mean?...whose hopes are we giving an account of...? A hope which is not the hope of the oppressed today is not hope for which I could give a theological account...we are batting the breeze and talking for our own self-satisfaction.[115]

Roberts Points to Failures

While happy to find a theological bridge between classical theology and the aspiration of black theology, Roberts did not hesitate to point out that the classical theologian who has managed to comprehend the relationship obtaining between the Gospel and liberation has failed to apply this knowledge to the evil of racism. Jurgen Moltmann, he points out, was correct in stating that, unlike the Lisbon Earthquake of 1775, the evil of Auschwitz was not evil in its natural form. Neither, he says, is racism. Roberts wrote that he was "highly suspicious" when Paul Lehmann, who, teaching close to the Harlem section of Manhattan focused on the liberation theology of Latin America rather than a political theology applied to the largest black ghetto in the world. Perhaps some statements and writings of Mr. Lehmann would nullify Roberts' personal comments, but one can appreciate the frustration on the part of the black religious community when politicians and theologians excoriate conditions in lands far removed, while appearing to lack passion for the continuing inability of black America to integrate into the body-politic in any meaningful way. As to Moltmann himself, Roberts recalled asking him about the value of his theology for oppressed people and not being satisfied with his answer.

Roberts points out that not only does the theology of hope not completely identify with black theology, it also does not completely identify with the theology of liberation which it inspired. He goes so far as to say that there are fundamental differences between the theologians of hope and liberation theologians, insisting that Moltmann and Pannenberg speak

---

115. Moltmann quoted in *Hope and the Future of Man*, ed., Ewert H. Cousins, Philadelphia: Fortress Press, 1972, pp. 55, 59.

out of the secure university settings of affluent countries, while Gutierrez devises his liberation theology from first-hand experience with suffering and oppression. And he goes further when he points out the differences between liberation and black theologies. First of all, the liberation theologians live in Latin America and, as all black theologians point out, have not lived the black experience. He expects that the liberation theologian will not take offence at this observation, for they themselves have pointed out that since European theologians have not had the Latin American experience, they cannot fully understand the oppression there. Other differences are that the nations of Latin America are Roman Catholic, underdeveloped, ruled by authoritarian governments and, as a result, heavily influenced by Marxism; their social task is to overthrow the existing social order. In contrast, writes Roberts, black theology is Protestant, and black theologians live in a developed, pluralistic land. He refers to the spiritual and academic freedom that enables the black theologians to arrive at truth unencumbered by any authority other than the truth itself and unmixed with a dogma to which findings and reflections must somehow conform. Black theology is a genuine political theology since it imports its dogma by direct involvement in the battle for liberation, without having to deal with obstacles presented by irrelevant dogmas and a hierarchy whose consciousness may not have been raised. Black theology, for Roberts, is theology at its best, and political theology at its freest. He agrees with Harvey Cox's observation that religions that retreat from a theological analysis of social evils in America face divine wrath.

Roberts does not reserve his observations to the white church alone, declaring that black theology has become ever more abstract and out of touch with the black church and its leadership. The black theologians have acquired plush professorships and reside in Ivy League Universities or ranking theological seminaries. Roberts does not excuse or paper-over defects in black theology that have come under his scrutiny. Speaking of his own religious development, he warns against the "by whatever means necessary" ethical norm that has been characteristic of black militants in their search for freedom and liberation. He writes:

> It is my view that liberation and reconciliation must be considered at the same time in relation to each other. The all or nothing, victory or death approach to race relations appears to be more rhetoric than reality, even to those who hold it.[116]

He does not excuse black theologians from adhering, strictly, to theological ethics on the grounds that white Christians are moral hypocrites. In fact it is, in his view, precisely because of the black theologians' neglect of ethical standards that the present impasse in race relations now exists.

Roberts: A Man of Reconciliation

Roberts extended his hand of reconciliation on many occasions. In contrast to Cone, for example, he claims that black theologians separated themselves from the white church primarily for social rather than theological reasons. Roberts suggests that black theology can learn from the theology of the Puritans and the Society of Friends, whose religious praxis are examples of how political theology has supported radical social transformation. The Puritans, he explains, believed that because God loved the world He inspires us to a social as well as personal salvation. Their political theology was a Covenant theology because they saw themselves as the exclusive people of God whose task was to do the work of God in history. The Bible was the blueprint for the organization of a church and state that acted as a vehicle for God's great design. Christians must be involved in the transformation of society. The Society of Friends, or Quakers, rejected the Calvinism of the Puritans but agreed that Christians must be involved in the social transformation, practicing the kind of "religion" of which Roberts approved. He concludes that the life-style of the Quakers pointed toward a new social order, a new Covenant, a new humanity. He wished he had found this kind of vision in Martin Luther King, Jr.:

> ...Unfortunately, the movement led by Dr. King had not given proper theological and ethical attention to power. Dr. King did not denounce power...He put all his confidence in *satyagrapha,* "truth force," a type of moral power, but he equated it with *agape*, love, rather than with the pushing and shoving of justice.[117]

Roberts criticizes King for not having analyzed and critiqued, from a theological perspective, the foreign entanglements of the C.I.A. and American-based multinational corporations, as well as American support for exploitative regimes in the Third World.

---

116. Roberts, *Liberation and Reconciliation* (Phila.: Westminster Press; no publishing date is given), p. 14.
117. Roberts, *A Black Political Theology*, p. 217.

Roberts' importance among his peers is reflected in his co-authorship with Cone, Wilmore, Roy Morrison and John Gatterwhite, of the Statement by the Theological Commission of the National Conference of Black Churchmen on "Black Theology," in 1976. His central theme, which he managed to hold onto with continued respect from his more militant black peers, is that there is a compatibility, a reconciliation between black theology and the universal Christian vision. This respect for Roberts was due, in part, to the fact that he early-on recognized the value and religious significance of acknowledging one's black skin saying that blackness is a fact of life which no Negro can escape, ignore or overcome. He calls on the black man to have, in Teillichian language, the courage to be black. For Roberts, this kind of acceptance of blackness is not the acceptance of resignation but an acceptance based on a convinced equality in terms of beauty, intelligence and goodness with the white man. Without this full and healthy acceptance, there would be the question (as posed by William R. Jones) as to whether or not God truly loves the black man.

Roberts: A Significant Theologian

Roberts has lived in the shadow of James Cone, but he has proven durable and significant in the field of black theology. With degrees from Edinburgh and further study in Cambridge, he was well-equipped to take up the task of scholarly commentary on the theology of the black church. Fellowships from the Ford and Lilly Foundations and the Association of Theological Scholars further equipped him for the serious studies which he undertook. When Roberts published *Liberation and Reconciliation* in 1971 as a critique of James Cone's theological program, it represented the first major attempt to take Cone to task. He presented a more inclusive perspective than did James Cone. His was not the spirit of confrontation. He saw liberation and reconciliation as the two main poles of black theology. He not only challenged Cone's right to dominate black theological thinking, he even said that theologians in general have been too generous to James Cone. His (Cone's) theological program has escaped constructive criticism, and he did not take this as a tribute either to Cone or to black theology. On the contrary, he stated that whites have used Cone as a "straw man" with the intention of rejecting black theology. By pointing to Cone's excesses they attempted to color all of black theology with the same brush. He calls for whites to join blacks in freeing the black community but he agrees with Cone that blacks should set the agenda for the liberation struggle. To those who see his movement for reconciliation as compromise, he responds that Christianity is rooted in the belief that "God was in Christ reconciling the world to himself." (II Cor., 5:19).

Roberts believes that a revolution in black/white relations would bring about a post-revolutionary spirit of reconciliation. Unlike Cone, he does not demand that whites become "black," in order for the reconciliation process to take its course, but that whites must begin the process by accepting blacks as equals. Roberts does not take the figure of the black Messiah in the literal sense, nor does he even approve of it. A black Christ, he says, deprives yellow and other-colored peoples a "Christ" in the Third World, and they need a Christological model of liberation as much as do blacks. In the black theologians' excitement over Pan-Africanism he cautions them that, exaggerated, it could lead to the same kind of racist exclusionism as that suffered by the black community. He points out that Asia and Latin America are also peopled with colored and oppressed populations; Africa has no monopoly on that score. Unlike Cone, who does not wish to talk about the afterlife until the black revolution is completed, Roberts reminds us that not only is our being grounded in God, but that the Christian life consists in reunion with the persons with whom we have become separated, namely, God and man. The truly liberated man is the man who has become reconciled.

William R. Jones

William R. Jones has posed the greatest challenge to black theology. A Unitarian-Universalist minister, graduate of Brown University (1968), former Professor at Yale Divinity and Professor at the University of North Carolina, Jones shocked a black community that was satisfied it had discovered "liberation" as the theological response to classical theology. From their own ranks a powerful adversary appeared who chided black theologians for having failed to deal with the very heart of racism, by neglecting to formulate a theodicy that was appropriate to an authentic theological analysis of racism. The black theologian, he said, had fallen into the trap of adopting uncritically the theodicy of classical theology, while living the illusion that he had made a radical change in theological methodology. In the process of trying to make sense out of the plight of black people in the light of the biblical message, black theologians claim to preserve the biblical truth that God sides with the poor and the marginalized. They see in the black historical experience and the biblical texts a symbiotic relationship, each a key to understanding the other. Jones intervenes and asks a disturbing question: "If black theologians' claim that God is on the side of the oppressed is true, why is it that there is no sign of this in the black Community?" Cone acknowledge the validity of the question and agrees that it has not been answered to anyone's satisfaction. He sees Jones' concerns as a breaking mechanism in the midst of runaway liberation rhetoric. Jones quotes from the writings of Albert Camus and Jean-Paul Sartre as the context in which he raises the question of suffering and he does this in a manner similar to the treatment of the subject in classical theology. Why does God permit so much evil in the world? Why does He permit any at all, since He is all-powerful? For Camus, classical theology's response to the problem of suffering and evil in the world makes a mockery of the Crucifixion and Resurrection in that these two major acts of Jesus' life are seen as giving example to others of patience in suffering. In this interpretation, Jesus is saying to us: "I suffered, so you can suffer also." Camus outlines the classical argument:

> Each time a solitary cry of rebellion against human suffering was uttered, the answer came in the form of an even more terrible suffering. In that Christ had suffered and had suffered involuntarily, suffering was no longer unjust.[118]

Cone Admits Contribution

Cone acknowledges Jones's contribution, pointing out that as an external critique from the vantage point of Black humanism, Jones' analysis has been and continues to be a challenge to Black theologians. Since Jones, he says, Black theologians can no longer be satisfied with a facile *internal* solution to the question of theodicy.

Cone does admit that suffering is the most serious contradiction of faith because the Gospel of God is proclaimed as the solution to human misery. It is precisely for this reason, Cone argues, that the mystery of suffering should be left as a question to which there can be no definitive explanation. The theological focus should therefore be placed on the political implications of our theological beliefs. It is in this area that strat-

---

118. Albert Camus, *The Rebel*, trans. by Anthony Bower (New York: Alfred A. Knopf, 1956), p. 32.

egies can be planned among Christians to alleviate suffering. Jones' analysis, he argues, can lead to despair, thus lacking that call to justice that can lead the victims of oppression to acts of liberation. If Jones, he says, can provide a theodicy which inspires the poor and oppressed to struggle against their oppression, then he (Cone) would support Jones' theological program. Jones challenged his brother theologians to prove their theory of the "suffering servant" as applied to the black race. Black theologians who are as far apart as Joseph Washington and Albert Cleage cite biblical support for black suffering. Cleage views black suffering as a logical consequence of having been chosen, in a special way, by God. To those blacks who say they have suffered enough, who wish to be liberated *now*, Cleage is not particularly encouraging:

> Some of you say, "If God is just, he ought to make it possible for us to go into the Promised Land right away. We have suffered enough." God cannot wipe out our weakness and faithlessness to each other.[119]

Jones finds the idea of planned suffering for blacks totally unacceptable, even if that suffering proves to be fruitful in the end. He cautions that black theologians cannot celebrate God's justice until they have established its existence. This is the part of the theodicy which he insists must be the first priority for any theology that wishes to call itself a theology of liberation. A contemporary black scholar in religious studies who began publishing his writing in the 1950's, Henry H. Mitchell, comments on Jones' theme that faith is sometimes, knowingly or unwittingly, used as a conceptual prop for oppression.[120] He believes that the answer to Jones' question, Is God a white racist? is clear. The black community, he says, *knew* that God was not pleased with slavery and they look forward with hope and joyful anticipation to their emancipation. This is why blacks have chosen the Exodus theme to represent their journey. Unlike Jones, he does not see current black theological thinking as tending to a form of quietism. In fact, he says, there has never been any quietism in the black community and it is likely that there never will be. He points out that there are innumerable accounts of the refusal of black preachers to desist from preaching even though they were repeatedly tortured and even killed for their actions. One need but note that the descendants of these fearless men

---

119. Albert Cleage, *The Black Messiah* (Sheed and Ward, 1969), p. 271.
120. Henry. H. Mitchell, *Black Belief* (Harper and Row, 1975), p. 118.

today continue to march in protest wherever and whenever black civil rights are violated.

Mitchell appears to be challenging Jones to prove that black history supports his concern, much in the same manner in which Jones had challenged the black theological community to prove that the history of black suffering supports their contention that God is on the side of the black community.

Jones first introduced the concept of the problematic nature of black theological theodicy in the *Harvard Theological Review*, in 1971. In this article he takes on Washington, Cone and Cleage claiming that, in spite of their strong rhetoric, the main question of black theology had not been answered, the question of the nature of God and His attitude toward the black community.

Cone quickly responded that Jesus' victory over suffering and death is proof sufficient for the belief that God sides with the oppressed and acts on their behalf. To Jones' demand for empirical evidence, Cone says that the problem of theodicy is never solved in a theoretical manner, but only within the context of one's faith. But Jones did not cease to press his message. He attacks the liberation models most popular among black theologians, those of Exodus (Cone) and the Diaspora (Shelby Rooks). According to the Exodus model God's insertion into the history of any particular people depends solely on divine election and not the supposed virtues of the people whom he favors. Cone views this divine election as freeing the black community from any pressures to be as good as any other people, for it would make no difference in their having been "elected."

Jones and the Exodus, Diaspora Models

Shelby Rooks (of the University of Chicago Divinity School) criticizes Cone's Exodus model saying that it does not relate in any historical manner to the real history of the black experience. He chooses to see the black community as a people in exile. He calls his model the "Diaspora Model." Jones does not take either model seriously. For him, neither the Exodus model nor the Diaspora model have substance. If God is active at all in the black community, His presence is, at best, ambiguous and thus irrelevant to an understanding of the meaning of the history of black suffering. Historically, the Exodus model has proven nothing, since the post-Exodus experience of the Jewish people has itself been full of persecution, suffering and humiliation. Jones expresses amazement that Cone would have chosen such a poor example, an example which might validate the opposite of what Cone is trying to prove. This is especially true since Cone has attacked the position of the theologian Emil Brunner. In Brunner's work *The Doctrine of Creation and Redemption* all that happens in life takes place within the knowledge and will of God:

> Cone wants to retain the sovereignty of God over human history...and yet avoid the damaging consequences of this affirmation in Brunner's thought. Cone seeks to accomplish this theological sleight of hand by restricting events of human history to those areas in which the oppressed are liberated.[121]

Jones agrees that joining the black historical experience to a multitude of biblical texts in a symbiotic relationship, each reinforcing the credibility of the other, is very attractive and appeals to a black oppressed community in desperate search for an answer to their plight. He argues, however, that the black theologians have not provided sufficient empirical evidence to back up their claims. He suggests that black theologians have not taken seriously the possibility of an evil deity. He writes:

> Clearly, in the context of Western monotheism, benevolence is as essential to the definition of God as is His existence. Hence there is an instinctive tendency to make God and goodness interchangeable terms and...to make either man or some other creature, e.g. the devil, the ultimate cause of evil.[122]

Theodicy, therefore, not black experience or any classical theological assumptions about the nature of God must be the theological starting point of a theology that wishes to deal with black oppression. He does not agree with Cone that Jesus' victory over death constitutes sufficient evidence for the belief that God is on the side of the oppressed. The Exodus model and the Diaspora model bear no proof that the black community is a chosen people. The Exodus was but a temporary relief from domination. For Jones, a reliance on an ambiguous God will lead the Afro-American to the same history of persecution as that experienced by the Jews. Jones approaches his theodicy from Sartrean principles that man is the sum of

---

121. Jones, p. 106.
122. Jones, p. 7.

his acts. We judge the character of a man by looking at his past. We deduce a man's present motives by what we understand is his modus operandi. There is no self which is radically different from the history of the man we know, and God's goodness or lack of it must be based on His past and present actions.

Whose Side is God On?

Jones emphasizes again and again that before a black theologian proposes that God lives in loving intimacy with the black community, that He is on their side, that He will deliver them and help them to "overcome," he must satisfy his reader that there is historical proof of these proposals. Does not the black experience suggest just the opposite? The option is to cling to the eschatological approach, where the black man is going to meet a just and caring God in the future. But this, he claims, would be a theological process which ignores the fact that in every other theological category God is judged according to His acts in history. The past and present condition of the black man is all that we have to go on. The future, by its very nature eludes us. Jones is now prepared to make some tentative conclusions of his own:

> It is easy now to see how the principle of man as the sum of his acts enlarges the contours of the concept of divine racism...if we do not come to our analysis of the divine nature with the presupposition of His intrinsic goodness...the question Is God a white racist? becomes an even more promising point of departure for black theology.[123]

For God to escape the accusation of being racist, He must provide for the black race what he has provided for the Israelites, namely a visible, tangible victory, an Exodus event. Even the greatest prophets called for this. Jeremiah asked, "Why is my pain perpetual, and my wound incurable?" (Jeremiah 15:18), and Jesus lamented, "My God, why hast thou forsaken me?" (Mark 15:34). For Jones, the eschatological option is a theological dead end, leaving the issue unresolved. His challenge to his fellow black theologians is radical since it questions whether the presumed theological radicalism of James Cone, J., DeOtis Roberts and others really does differ, substantively, from classical theology. Jones calls for black theology to make a radical reassessment of its theodicy. Until

---

123. Jones, p. 11.

such an assessment is made, he prefers to consider himself a "brother" to those who call themselves humanists.

Summary

Each of the theologians we have considered has attempted to analyze the nature of the Gospel of Jesus Christ in the light of oppressed black people, so that they can make a connection between their humiliated condition and the Gospel hope that freedom can be attained either because God is on their side or because, as in William R. Jones' view, the black church has been a source of education and inspiration even if it has not been precisely correct in all the judgements it has made relative to the presence of God in their community. Each of these men has attempted to fill in what he sees as large gaps created by white theology in the understanding of the real meaning of the Gospel message and what constitutes being a Christian. They are intellectual pioneers, since the literature available to them needed forming and shaping as academic theological literature. They had to rely on reading each other, for the most part, with the full knowledge that they were creating a whole new body of theological literature without the rich theological sources available to the classical theologian, and they had to do this in an atmosphere already charged by the militant actions of their contemporary black power brothers and sisters. They are, in my opinion, the greatest living theologians, the ones who most succinctly and passionately articulate the heart and guts of the gospel of Jesus. Only the Liberation Theologians of Latin America approach them in analytical scope and grandeur.

CHAPTER FIVE

A CRITIQUE OF BLACK THEOLOGY

Black Theology, like Latin American Liberation theology, developed within the context of a modern understanding of the relationship between religion and liberation. For this reason, a critique of such a theology necessarily moves beyond the traditional focus in theology on revelation and tradition as starting points in theological reflection and investigation. In saying this, I am in no way implying that the problems or events which preoccupy much of classical theology were not, at the time that reflection on them began, contemporary. What I am saying is that to the extent that such problems and events are now historical and handed-down, they take a less prominent position relative to the focus of liberation theologies, including black theology. Here, I shall critique black theology not only from the point of view of revealed truth in the sense of the "deposit of faith," but also as the unfolding of revealed truth as the black theologian consciously attempts to apply to today's world of the black diaspora the kernel of truth imbedded in the Gospel message. A critique of black theology involves the facts and questions derived from recent black socio-economic history, and an understanding that divine truth continues to be revealed as the black theologian reflects on the relationship that exists between God and this particular people with a history of slavery and oppression.

This critique will also be done with a conscious effort to understand that the black theologian is both moving into areas of theological investigation not covered by his classical colleagues,[124] and that he comes to this theological project with very strong emotional reactions to slavery, racism, economic exploitation and the experience of having shared, for the most part, in the neglect experienced by the black community in general. It is also critically important to understand that the objective situa-

124. Black theologians tend to use the term "white theology" indiscriminately, but for the most part they are referring to Classical Theology.

tion of black theology is that it contains within itself partiality, uncertainty and undeveloped as well as underdeveloped theological criteria. The critique will include observations on the "partiality" and "universality" of the content of black theology, as well as the importance of social versus personal sin, and the term "black" in black theology. The contributions of black theology to theological pluralism in America will be considered, along with the academic and religious limitations to a theology that has "anger" as a component and is flawed by an elitist, self-imposed isolation.

Historical and Liberating Praxis

Gutierrez makes an important distinction between historical and liberating praxis, and this distinction is key to an understanding of all liberation theologies, including black theology. Gutierrez defines historical praxis as the mechanisms by which the transformation of the natural and social conditions of humanity takes place, the actualizing of the ability that humanity has to create and fashion a society. Liberating praxis, on the other hand, is the effort to fashion historical praxis in such a manner as to achieve conditions of social, economic, cultural and racial justice. In the black community, historical praxis has been replete with unjust structures that cry out for a liberating praxis. Just as the methodology of liberation theology leads to a prior option for the poor and oppressed and a commitment to their liberation, so black theology's prior option is for the black poor and the black oppressed. Black theology, as a form of liberation theology, consists in a "re-reading" of the Gospel message from within the context of liberation praxis.

Liberating praxis in the black community is a direct product of a raised consciousness in black men and women that they are the agents of their own history and that they bear full responsibility for the social and economic structures of living together in an imperfect world. To a great extent, this consciousness-raising can be attributed to the work of Martin Luther King, Jr., and it has been furthered by the work of the black theologian, who sees himself as an agent of consciousness-raising in the black community, with a view to encouraging black people to take their future history into their own hands. Blacks are encouraged to create for themselves structures of justice, even in the context of a dominant white society.

Black Theology's Impact on the American Conscience

Perhaps the most valuable by-product of black theology's emphasis on social and racial justice is the impact this has had on the conscience of white America, even with reference to its perception of justice within its own ranks. Black theology's success in highlighting the moral dimensions of racism and economic inequality has made an enormous contribution in supplying for the inability of white American Christianity to critique the outstanding moral issues in American society. A white Christianity which was blind to the deeply imbedded racism in its body-politic and which had difficulty perceiving the injustice of America's military involvement in nations of "color," has learned much from perceptions emanating from the black community. These perceptions are mediated through the black church and articulated most clearly by the black theologian. The first judgement about black theology is that it has been a noble attempt to bring about justice where justice has been illusive. It is time for white America to acknowledge that its religiosity and its theology is far inferior in terms of the gospel message to that articulated by the black community. It was not the white clergy and their bishops in this nation who marched for freedom and explained in what its authentic existence consists; it was the poor black preacher and his followers who, enlightened by faith, and strengthened in the long night of suffering and humiliation, issued the clarion calls for justice. They, and only they, of all the people of religion in our nation, had the righteousness to right the moral wrongs. Black theology can be credited with making a significant contribution in raising the social consciousness of all American religious life by pointing out the deep split that has traditionally existed in classical theology between an individual and social interpretation of the Christian religion. The black religious leaders who led their people in demonstrations singing "We Shall Overcome," were giving public testimony to their theological belief that individual salvation can only be accomplished within the context of social liberation, a belief minimized in the context of classical theology, with terrible consequences, the most prominent of which in America being a deep-seated racism, both inside and outside of the white church, of whatever denomination. What the black religious leaders hoped to "overcome" first of all was the intransigent attitude amongst white Christians toward letting their black brothers "go." If individual salvation depends on one's attitude toward social salvation, then it is clear to the black theologian that white theology has mediated a false opposition between individual and social salvation. This is the deepest substantive question of theology today. John Carey, professor of religion at Florida State University, believes that black theology has become not only a rallying point for the black Christian community but the conscience of white America. Carey approves of Robert McAffee Brown's paraphrase of Karl Marx that theologians in the past have tried to understand the world rather than change it. Black theology has not hesitated to learn from existentialist philosophy the importance of the authentic individual, but has moved beyond that emphasis since it tended to create an apolitical individualism that did damage to the socio-political dimension of prophetic Christianity. Cornel West explains that the prophetic Christianity of black theology recognizes and retains the concept of freedom as both existential and social. For West, existential freedom anticipates history by empowering people to fight for a social freedom which is a matter of this-worldly human liberation. Cone agrees:

> The focus on praxis for the purpose of societal change is what distinguishes Marx from Hegel and liberation theology from other theologies of freedom.[125]

West characterizes black Christianity as having four central elements: the philosophical methodology of dialectical historicism, the theological stance of prophetic Christianity, the cultural inheritance of Afro-American humanism and the social theology and praxis of progressive Marxism. That black theology has seen itself in need of serious dialogue with Marxism is one more negative effect of the inability of white and black theologians to find enough common ground to enter into serious dialogue with each other on the need for social change, and the strategies for bringing it about. The black community has moved on its own, opting to receive its political leadership still, as it always has, from its clergy. Martin Luther King, Jr., Abernathy and Jesse Jackson are a few of the more recent examples.

Pluralism

Black theology has both created and encouraged new insights into the religious dimension of pluralism in American life, pluralism that calls for an attitude of respect and appreciation for the various communities and

---

125. Cone, "Christian Faith and Political Praxis," in *The Challenge of Liberation Theology*, p. 60.

cultures which make up this society. Black theology is both a result of the need for pluralism and a sign of the extent to which classical theology has failed to lead the way to its realization. There has been no incorporation of the black community into the body of American Christianity, and although black theology's critique is at times rhetorically exaggerated, the indictment that white Christianity has failed to incarnate itself in the black community rings true. Black theology, therefore, by its incisive and thoughtful (even if angry) criticism has contributed to the advancement of a truly humane American pluralism. America's theoretical boast of possessing the pluralistic society, to the extent to which it has been actualized, has become incarnate in America by means of the black community, not by its founders, not by its dominant class and race. The *Statement by the National Committee of Black Churchmen*, June 13, 1969, addressed itself to the origins of a separate black theology:

> This indigenous theological formulation of faith emerged from the stark need of the fragmented black community to affirm itself as a part of the Kingdom of God. White theology sustained the American slave system and negated the humanity of blacks....

An Important Contribution

Perhaps the most important contribution of black theology is its conclusion, arrived at independently of the Latin American liberation theologians, that theology is not the first act, but the second. This understanding goes to the root of how theology is done. The first act, in liberation theology, is the commitment to the poor and the oppressed. Theology without personal commitment rings hollow no matter how erudite the discourse. The theological meaning of praxis discovered during theological reflection is absolutely necessary, but praxis comes before theology. Gutierrez laid the foundation for this understanding by explaining that theology (the second step) follows praxis.[126] James Cone, in answering the question: How do we do theology? agrees that praxis, reflective political action that includes cultural affirmation, comes before theological reflection. He insists that black theology does not believe that revelation is a list of fixed doctrines, that there is no truth outside of the concrete historical events in

---

126. "What Hegel used to say about philosophy can likewise be applied to theology; it rises only at sundown. The pastoral activity of the Church does not flow as a conclusion from theological activity rather, it reflects upon it...." A Theology of Liberation, p. 11.

which people are engaged. Its focus on religious and social analyses distinguishes it from the abstract classical theologies of Europe and North America. Commitment to the poor changes the focus of the reason for doing theology, since it arises from the experience of evil in the socio-political structures that oppress and impoverish, thus implanting the Gospel message in the academic as well as socio-political life of a nation. Theology that follows upon experience is a critical reflection upon our commitment to mold a world in which Gospel values will be pervasive. Commitment is act one, and reflection is act two. Doing and saying are inextricably entwined so that what one says is what one is doing. Black theology joins liberation theology in proclaiming that orthopraxis, not orthodoxy, is its criterion. Theology thus becomes a reflection on the reality of a God who is in active solidarity with those struggling to overcome cultural, economic and political oppression.

Black Theology and Metanoia

Black theology has been and continues to be a valuable instrument for self-critical, historical reflection on the part of classical theology. If Cone's invitation to white theology to "become black" leads to a practical metanoia in which it accepts guilt on behalf of the Christian community for historical crimes committed against blacks, then the beginnings of a genuine solidarity with the exploited peoples of the Third World becomes possible. The invitation to white theologians to become "black" is a socio-theological designation, not a chromatic one. If his invitation is misunderstood, it can be misread as an affirmation of a reverse racist ideology. The burden is on James Cone to make it clear that this is not the case. Paul Lehmann, professor emeritus of systematic theology at Union Theological Seminary, in New York, recommends that white theologians listen to Cone, agreeing to the characterization of classical theology as chauvinistic and parochial.

That black theology has become a catalyst to white theology gives it a value which is difficult to overestimate. As a theology which does not represent the dominant culture, its ability to become the "liberator" of American and European theological parochialism testifies to the moral authority and integrity of its content. The *metanoia* of which black theology speaks is not limited to sorrow for past sins; it involves rejection of an historically debilitating past, and the courageous assumption of a new and radically different path. This new path is not satisfied to use moral persuasion in an attempt to bring about racial and social justice; it envisages questioning the very economic and social foundations of American society. James Cone is tired of using moral suasion and wants to move on to social analysis. He despairs of the former black theological stance that by pointing out the contradiction of a Christianity riddled with racism substantive changes in classical theology would take place; he now concludes that this hope was born of a misplaced trust. At the height of Martin Luther King, Jr.'s influence, the American black leadership reflected his views that the appeal to the moral goodness of the white community, in a spirit of integration, was the path that should be followed. The growing influence of Malcolm X, however, produced a shift away from moral suasion to confrontation and demands that the rights of blacks be respected by the white community. Black theology played a key role in bringing these two approaches together in its acceptance of the need for social analysis within the context of Christian biblical themes. There is no doubt that the rhetoric of black theology has made theological dialogue with the white community difficult. Simultaneously, however, it also played an important role in containing the extreme militantism of the Black Power movement, absorbing some of the rhetoric of Malcolm X in a theological blend softened by the non-violence of Gandhi/King.

A Few Notes on Malcolm

Malcolm X has made a spectacular comeback. In Trinidad long lines of expectant youth waited impatiently to purchase tickets to the movie that will tell them all about the legendary black hero assassinated before they were born. They wore T-shirts emblazoned with his image and his searing messages; they are looking for leadership, even from the grave, even from beyond their borders. What did Malcolm say? what did he represent? They don't really know; they just have that *feeling*. I can relate to the feeling. I experienced it when Father Dan Berrigan spent a night at my home in Newport. It wasn't so much what Dan said; I'd heard it all before. It was his presence. It was the kind of thing that happens when Dan Berrigan is around. He gave a talk to my Naval War College students, after which one of them, an experienced aviator, asked to chat with him. Two days later, in class, the aviator told me: "After listening to Berrigan, I've decided to quit the service. I love my country and it has to be defended, but since that talk I really don't want to drop another bomb, even on an enemy." That same student had said to me when I announced that Berrigan would address his class: "We'll make mince meat out of that anti-American."

Like all heroes, Malcolm suffers from the tendency of admirers to freeze that moment of his message that appeals to them. Some want to be arm-chair revolutionaries; others want to burn buildings; some just want to read more about him and incorporate the new knowledge into their lives. He is all things to all the phantasies that momentarily remove his fans from the ugly realities of the ghetto, economic or mental. Malcolm, of course, was not tied down to any ideological position; revolution for him was the struggle for freedom as carried out in any form - like refusing to sit in the back of a bus. Ironically, among his brother leaders in the black movement he emerges as the wild-eyed idealist; in reality, he was the most pragmatic of them all. Martin Luther King gave the same speech to black and white alike, to the University student and the ghetto child; he was, in a sort of mystical sense, above the fray that whirled all about him. Malcolm focused on the particular suffering standing in the crowd in front of him. He became molded by the configuration of the crowd, *this* crowd. He was of the genre of Stokely Carmichale, Amiri Baraka and Floyd McKissick, names little-known in the white community because, unlike Martin Luther King, they spoke with a uniquely black voice, from the guts, using the language inherited from hard-working, street-walking relatives and friends. His "Message to the Grass Roots" delivered at the Detroit Freedom March in 1963 evoked personal responses from the crowd of more than two hundred thousand people. Martin Luther King, perhaps, did not understand as fully as did Malcolm that black people are locked in a struggle against the white community. It sounds fiery to say this, but it's true. To struggle is to stand on the front lines as an equal with the adversary; to work hard at reconciliation could indicate an admission, not of humility, but of inferiority. Malcolm just stood there, toe-to-toe in what was, in fact, an unequal struggle, a David and Goliath scenario. He just flat-out declared that Black people are oppressed *because* they are Black, period. He preached that if blackness is the reason why the community is oppressed, then blackness must be the reason why the community organizes. And he further delineates the struggle when he explains that the enemy is not so much the white man, as the white power establishment, an institution as old as the Republic. He made it clear that insti-

FIGURE 6: Infamous for its brothels, Anthony Street (renamed Worth in 1854), is shown here in the mid-1860s from the Five Points intersection facing west. This block, between Orange and Centre streets, was one of the two on which the majority of Lansdowne immigrants settled. In the building at the corner on the left, Bridget Mangin operated a well-known bordello. Andrew Crown's grocery and saloon occupied the ground floor. Courtesy of the New-York Historical Society.

FROM ALL APPEARANCES, THEN, the Lansdowne immigrants must have been some of the most impoverished residents of Five Points, a neighborhood renowned for its destitution. After all, the Lansdowne men were especially likely to be lowly paid menial laborers, and they lived with their families in the district's most rundown tenements. The Lansdowne immigrants' propensity to open accounts at the Emigrant Savings Bank makes it possible to compare these appearances with reality. Of the 12,500 accounts opened at the bank in its first six years of operation (from September 1850 to August 1856), 153 were opened by Five Points Lansdowne immigrants. Although it is difficult to determine what proportion of Lansdowne immigrants had accounts, it appears that about 50 percent of the Lansdowne

tutionalized power is the enemy. Moral white men are corrupted by the power of the institution; they are overcome by the power of it all. Had he not pointed this out, he might have misled millions of his followers world-wide; he might have used emotionalism to create yet another psychological scam aimed at self-glorification. His perceptions created space in which other black intellectuals could move within his orbit, making their own contributions—the economists and sociologists, the philosophers. All found a home beneath his all-embracing cloak. As for the white man, after Malcolm he could no longer comfort himself with Martin Luther King's dream (appealing as it is) of white and black children playing together on each others lawns. Malcolm pointed out that American industry needs a permanent underclass, the exploitation of which provides the kind of profits used to purchase corporate jets and multi-million dollar vacation homes. He knew that labor, and labor alone (whether intellectual or physical) creates wealth; he knew that the industrial giants understand this very clearly. The net result of such knowledge is that one never strategizes to "cooperate" with the captains of industry because, their rhetoric notwithstanding, they are not going to vote themselves out of existence.

There was a touch of Martin in Malcolm. He believed that by addressing himself to the international community, by appealing to their sense of morality, the non-American white conscience would respond, would apply pressure. It didn't happen. He miscalculated the intentions of non-American whites. He thought they would be different. He believed he could mobilize international black support, indict racist America before the non-white world. He got plenty of rhetoric, but international affairs carried on much as before. At the time of his death he may have already begun to despair of a Christian solution; he contemplated the separatist option.

Cone, in his book, *For My People*, regrets that black theologians did not spend less time listening to Martin L. King, Jr. than to Malcolm X, since Malcolm appealed not only to black religious and political leaders but also to young blacks world-wide. In spite of criticisms from such eminent Afro-Americans as Carl Rowan and Thurgood Marshall, Malcolm, writes James Baldwin, is a man for all seasons, for every persuasion; all claimed him. The man who called himself Malik El Shabazz and made pilgrimages to Mecca was, in Jesse Jackson's words, not finished yet. Maurice Bishop and Kendrick Radix of Grenada, Tim Hector of Antigua, Rosie Douglas of Dominica, Ralph Gonsalves of St. Vincent, Geddes

Granger and Yusin Abu Bakr of Trinidad and Tobago caught his illuminating fire, absorbed it within their own politico-cultural milieu and passed the torch to their followers. He was Ossie Davis' "shining black."

CHAPTER SIX

PROBLEMS TO BE DEALT WITH

The Rhetoric of Black Theology

Black theology has its own set of code words that make it not easily assimilable to the white theological community. It speaks of "black logic," a "black God," a "black Virgin." This has contributed to black theology's being tolerated rather than accepted in a society which prides itself in academic toleration and accommodation. This is unfortunate and perhaps could have been avoided if its birth as an academic discipline had not been accompanied by the frenzied rhetoric of the black power movement. Histrionics became a feature of its presentation. What it gained in the glare of publicity attendant on such social chaos as the Watts riots, it may have lost in authority and, perhaps, quality. Black theology has in it a strain of puritanism that demands all-or-nothing imperatives, a reluctance to compromise, a desire to judge everything by a single standard neither agreed to nor accepted by a large cross-section of the American academic or religious community. Its passion sometimes overrides everything else; the black theologian cannot expect that the religious concerns of those who do not share the history of the black community will suddenly be transformed into a sudden surge to articulate black theological premises.

Paul Holmer, of Yale Divinity School, in commenting on the writings of James Cone, says:

> ...Black theology, as part of the social movements and the new academic fever, acquired a certain importance simply because it was exclusive, shrill and demanding. It fed the need for drama and histrionics...For a while it gave rationale to the notion that gladiatorial talk was going to get an equitous moral order finally established.[127]

---

127. Paul Holmer, "About Black Theology," Lutheran Quarterly, 28, No. 3 (1976), p. 33.

The black theological community must resolve the struggle within its own ranks relative to the importance of being "black." Major Jones and DeOtis Roberts favor emphasis on personhood. Cone, in contrast, equates the denial of blackness with the loss of identity and the foundation for sin in the black community: "Sin then, for Black people, is the loss of identity."[128] The differences in these two perspectives can be attributed to Cone's attempt to develop a model for black theology, while Jones and Roberts are seeking a model that gives theological substance to what Christian love demands in the context of race relations. Further, Cone's militant stance vis-a-vis black power is not shared by Roberts who asserted that many blacks who are not Christian are associated with the aims of black power, concluding that it is dangerous, therefore, to equate the two (black power and Christianity).

Black/white theological dialogue would be advanced if the black power theologian makes an effort to explain that the word "black" has a double meaning, one physical the other ontological. He must explain that black does not, in black theology, refer to a people whose skin color is black but to all who are in solidarity with the oppressed, while not conceding that white theology has succeeded. And if the explanation of the term "black" is done within the context of dialectical materialism, then the black theologian moves away from the use of symbols that are derived from the black church experience. The influence of his theology on the faithful can only suffer from such a diversion unless there has been a good deal of intellectual preparation, the kind of preparation that was done in Latin America, for example, by the work and writings of Paolo Friere.

Reaction to White Racism

The theological agenda in a given time and place is nuanced by the socio-economic atmosphere in which it lives. The post-Reformation emphasis on doctrinal orthodoxy is perfectly understandable, given the disintegration of what had been (in Western Christendom) one theological fabric. It was, in part, as is much of theological history, a reaction to events of the time. The strength of such theological reaction lies in the willingness of a theological community to come to grips with contemporary issues. The weakness derives from the fact that reaction to events tends to be negative relative to the problem under consideration. A theology that is based on a

128. Cone, *A Black Theology of Liberation*, p. 196.

reaction to a perceived threat may be necessary historically but loses much of its creativity and credibility. The new black theological consciousness that began with the publication of the "Black Power Statement" of July 31, 1966, and the writings of James Cone and Gayraud Wilmore were a theological reaction to racism as found in white society in general, and the white churches in particular. James Cone, himself, after contributing to this kind of negative theology, had this reflection:

> A similar weakness is found in many texts on black theology...It was as if the sole basis for black theology were racism among whites. But if so, and if racism were eliminated, then there would be no need for a theology based on the history and culture of blacks.[129]

The emphasis on angry reaction aborted the development of an ideology that would transcend blackness and would contain prophetic utterances which would provide the foundation for the resolution of the problem of white racism and advance a strategy for black integration into the white American body-politic. The rhetoric of reaction was not followed-up by a careful theological analysis which might have appealed to those members of the white theological community who were sensitive to the pain, frustration and anger contained in the black theological rhetoric.

The black theological community, also, did not then and has not since developed a well-organized and consistent program to educate and motivate the black churches. The National Council of Black Churches itself was supported by money from white institutions because it had not developed a strategy for enlisting moral and financial support from the very black community in whose name it claimed to speak.

Limitations of Political Theology

Black theology was and continues to be, in part, a response to socio-political events taking place in the black community, or in the white community in so far as they impact on the black community. Attendant on such a theological response is the reality that particular events, or series of events, can easily become dated or irrelevant, given a new event or series of events which contradict or greatly modify the reasons for the original response. Black theology, like other liberation theologies, defines itself as evolving out of the events of the place where it is practiced and in the time

---

129. Cone, *For My People*, p. 87.

frame of those events. By self-definition, therefore, it is limited and circumscribed by place and time. Its claim to universality lies in its conviction that local solutions and reflections have universal applications; this claim is problematic if it hopes for an emotional/motivated response from communities living on the edge, or further, of particular conflictual events.

Religious Context Should be Maintained

Black theology shares both the gains and losses inevitably incurred in the process of establishing a political theology, which black theology avowedly is. The gains, as I have pointed out, include the development of a badly-needed black perspective on theology, a raising of consciousness, and a new sense of both religious and socioeconomic direction. The losses are only slowly beginning to emerge, as black theology and non-black theology reflect on the years since the emergence of black power and its theological counter-part. If Christian hymn singing and Christian preaching take place in a context of socioeconomic conflict often enough, there is the real danger that praising God and preaching the core message of the Gospel gets lost in the struggle to attain social equality. If the reasons for gathering together in a religious context are continually confined to non-religious matters, then, eventually, the religious context will become associated with and even identified with non-religious matters leading, eventually, to a situation in which the religious context could lose its identify. If this happens, then the circle is completed and the moral base for socioeconomic justice can also erode. The question of socioeconomic justice is not separated from the liberating grace conferred on man through Jesus Christ. The introduction to the Vatican Document on liberation theology puts the matter in context:

> The words of Jesus, "The truth will make you free," must enlighten and guide all theological reflection and pastoral decisions...Thus the quest for freedom and the aspiration to liberation...have their first source in the Christian heritage.[130]

The black theologian needs to be conscious of and acknowledge the manner in which extreme right wing and extreme left wing groups and

130. "Instruction on Christian Freedom and Liberation," issued in Rome on April 6, 1986, by the Vatican's Congregation for the Doctrine of the Faith." Selections quoted in *The New York Times*, Sunday, July 6, 1986, p. 14.

movements eventually use religion as a political tool. The transformation of the Reverend Jerry Falwell religious movement into an openly political movement is a case in point. To practice black theology within the context of a specific political movement would be to exploit the uninformed and unsuspecting faithful who (especially in the black community) are excited by the spectacle of a religion and a theology that critically reflects on racism, economic deprivation and the arrogant domination of one cultural ethos by a another.

The Question of Means

James Cone's determination to achieve black liberation "by whatever means necessary" implies an openness to the use of violence; not violence as last resort, but simply because it might be useful and is available in the struggle. To the extent that Cone dominates the black theological discourse it is most unfortunate that in his passion for the black cause he did not take the time to clarify this determination in the light of Christian principles. How does his attitude differ, for example, from that of the Black Panther Party, or the militant wing of the Black Muslims? The question of "means" is especially important today in view of the fact that much of the world's nations are sustained by the threat of increasingly perfectionized violence. It is true that nations, when gathered together in international fora, condemn the use of torture and advocate the humanization of the penal code, but this progress is more than compensated for by the spread of torture and the proliferation of modern military technology in our century. For a theologian to advocate the liberation of a particular people "by whatever means necessary" is tantamount to abandoning Christian principles and rationalizing the use of force in the realization of political goals. It is one thing to say, "Revolution is justified," and quite another to say, "Let's make a revolution; it's the easiest way to solve the problem." Cone would serve his own theological reflection well if he stressed that violence should be limited to the last-resort scenario and that, even in that context, the intention must be there to limit the violence as much as is humanly possible. That Cone does not make the effort to do this is disappointing.

Black Theological Revisionism

The black theologian looks at American history through the eyes of a black community which does not have a prominent presence in that history. The black theologian is looking for what Donald W. Shriver of Union Theological Seminary calls "the missing documents."[131] American history read from the perspective of a Sojourner Truth or a Harriet Tubman, even a Malcolm X, will give students of that history perspectives that, in some cases, may differ substantially from that of American history written by white academics. After 250 years, black historians and theologians are finally revising American history from their perspective. To the extent that the product of their search is a more honest and complete study of the American past, such revision must be welcomed. But, the re-writing of history is also fraught with danger. For one thing, the revisionist, because he is writing history from a particular perspective, can easily fall into the same traps that made the need for revisions necessary in the first place. If the black theologian presents us with some of the "missing chapters" of American religious history, he must avoid as distortion of white religious history that will provoke further revision to straighten-out new distortions. Donald Shriver comments:

> Church historians, theologians and ethicists will not serve the American church or people if they systematically ignore the possibility that there is something good, as well as something evil, hidden or revealed in American history...That so many (blacks) should be so willing to "keep on keeping on" in this racist society constitutes an enormous compliment and hope for our society.[132]

Since black theological revisionism is subject to the temptation of interpreting Scripture in such a manner as to tailor it to the needs of the black community, the words of James Cone in this regard are worthy of reflection:

> If black religion is identical as the *only* possible interpretation of the Bible for black people, then what is the universality implied in the particularity of black religion?...To be Christian and human means developing a perspective on life that includes all peoples.[133]

Black theology's attempt to expose the particular ideology of classical theology is the more credible the more it avoids the quicksand of an ad

---

131. Donald Shriver, "The Churches and the Future of Racism," *Theology Today*, Vol. 38, No. 2, July, 1981, p. 153.
132. Schriver, p. 154.
133. Cone, "Epilogue: An Interpretation of the Debate Among Black Theologians," in *Black Theology: A Documentary History*, p. 619.

hoc ethics which skirts around the validity of accepted Christian principles. If black theology fails to transcend particular social interests, then how will it be able to distinguish the ethical character of one even from another?

The Importance of Communication

Black theology, with the power of its just cause ever at hand, could make a greater effort to communicate in a language that others can understand. The culture that underpins black theology is remote and to some extent unreal to the very white academic community so much in need of understanding the message. In the effort to communicate under such conditions, symbols and words undergo changes in meaning and definition. Such transformations can have a net effect of causing confusion and outrage rather than acting as vehicles that inform and motivate changes in attitude. If the goal of a message is simply to "tell somebody off" that is one thing; if the goal is comprehension leading to dialogue, that is quite another. Tell me off if you want; its likely I deserve it. Then, let's get down to business. If it is true that, in the words of C. Eric Lincoln, "White Western Theology has contributed significantly to the involuntary invisibility of black people,"[134] it is also true that rhetoric emanating from the black theologian might contribute to an unintelligibility that lessens the possibility of a fruitful dialogue that could lead to visibility.

In the process of being the negation of an historical negative, black theology serves its own interest by being perceived as more Christian than the "Christianity" it challenges. But if it turns out that black theology has "become" a sanitized version of the Black Muslim movement then, valuable as the Black Muslim movement has been to the black community, black theology negates itself as Christian theology by employing a rhetoric that is rooted in Islam's historical advocacy of violence to attain its religious ends. Lincoln writes:

> If black theology is to be anything more than a counter-myth and a more extravagant fiction, it must avoid at all costs the disparagement of principle and the notion that the whole will of God is discoverable in one momentary event or one set of principles.[135]

---

134. C. Eric Lincoln, "A Perspective on James H. Cone's Black Theology," Vol. 31, 1, Fall, 1975. (*Union Seminary Quarterly Review*).
135. Ibid, p. 22.

A black theology that becomes just another exclusive theory claiming truth for itself militates against Christian openness and undermines whatever moral authority it attained in its powerful and justified critique of classical theology. Black theology might do well to invite a response to its thoughtful and critical discourse. White and black alike face the ambiguity of shared social existence and both theological disciplines must take seriously the unique American pluralistic experience if we are to move forward to a more creative society. Rather than despair at the different paths our pluralistic theologies take us and attempt to sanctify and codify each path as the "true" path, the opportunity exists for the theological community to make an important contribution to the humane pluralism of which our official societal institutions and spokesmen speak so proudly.

Pessimism In Cone

There is a pessimism in Cone that I find discouraging. He rightly calls upon the black community to revolt against racist structures, but precludes a happy outcome to such a struggle. The struggle, he says, must take place because such action is in harmony with the revealed activity of God, not because there is hope of a joyful outcome. I agree with Cone that one must not struggle against evil only when there is a certainty of victory, but surely a future does lie before us in which the combined efforts of many disciplines, groups and dialogue can lead to a gradual destruction of racism in America. Cone appears trapped in what has been, despairing that the future could be different. The black theologians complain that the Theology of Hope is too narrowly eschatological. It seems to me that it is precisely at this point that an opportunity presents itself to work for a hope that is realizable on earth. Maybe Cone is right, but if he is, that would bring about another entire debate, it seems to me, about the efficacy of the very Christianity which Cone preaches so eloquently. In any case, there is very little evidence that Cone's pessimism is shared by black worshipers, for their prayers are replete with references to the "Exodus" as the paradigm of their own coming liberation. When Cone says: "Black theology cannot accept the view of God which does not represent him as being for blacks and thus against whites,"[136] he places himself outside of the black church tradition. He loses the balance which resides in a black

136. Cone, A Black Theology of Liberation, p. 131.

community which is far more forgiving and hopeful than some of their theological representatives.

Black Theology and Feminism

The liberation insights articulated by black theology have not been consistently applied to the area of women's rights. And even when the black theologian has made a tentative move in that direction, he has come up against the attitude of most male black preachers, who consider women's liberation a creation of the white women's movement. Cone observes:

> ...most black males in the ministry labelled women's liberation as a "white thing." I remember my reluctance to even consider the issue, and I am greatly embarrassed by my silence and by the sexist language of my early works on black theology.[137]

Since black theology calls itself revolutionary and proclaims that it is working for liberation in the widest possible context, its hesitation to address issues of black women's rights leaves it open to the suggestion that what it has criticized in classical theology it itself has practiced. Cone admits the problem:

> It is truly amazing that many black male ministers, young and old, can hear the message of liberation in the gospel when related to racism but remain deaf to a similar message in the context of feminism.[138]

This admission makes less tenable the presumption that the partiality of black theology is legitimate since it has universal applications. Black theologians have interpreted American society in terms of slavery and racism. This kind of analysis should alert them to the fact that the "place" of women, as defined in that kind of society, might well be in need of revision and reflection. Perhaps, as black feminists suspect, theology done by men has served to maintain the patriarchal structures in American society. Black women have suffered from the same fate as white women in that black men have defined the nature of their own sex in terms of reason and intellect and that of women in terms of intuition and emotionalism. Jacqueline Grant, a lecturer at Harvard Divinity School in the late 1970's, wrote:

---

137. Cone, *For My People*, p. 97.
138. Cone, "Black Ecumenism and the Liberation Struggle," given at a seminar held at Yale University, Feb. 16-17, 1978.

> Black women are enraged as they listen to "liberated".black men speak about the "place of women"... in phrases similar to those of the very white oppressors they condemn....Inasmuch as Black men have accepted the sexual dualism of the dominant culture... Black theology must subject the Black Church to a "self-test."[139]

In addition to the patriarchs (like Cone, Washington and Williams), African American women theologians have also had a significant impact upon the development of black theology. Matriarchs such as Jacquelyn Grant, Delores Williams, and Katie Cannon have proved to be major commentators on the persistent sexism displayed among both ministers and theologians of the black church.[140] Black theology's record on the amount of energy it has expended on women's rights can have some positive side-effects, if reflection on this deficiency results in a reduced confidence in the infallibility of its pronouncements and a less strident righteousness.

Lack of Social and Economic Analysis

Black theologians have been very slow to use the tools of the social sciences in an analysis of the social, political and economic structures that dehumanize the poor. It has offered no alternative to an economic system which has not benefitted the vast majority in the black community. So strong in the late 70's and early 80's black theology has lost some of its vitality and credibility by failing to advance solutions to the very socioeconomic problems which it proclaimed as totally unacceptable. Major Jones recognized the problem and called on black theology to develop a new correlation between the eschatological origins of Christianity and the present-day economic and sociological factors in society that seek a better future for the black man. The failure to come to grips with a thoughtful socioeconomic analysis is due, in part, to the naive internalization on the part of the black theological community of the ideal contained in the "American Dream." There has not been a sophisticated understanding that if the affluence of the white community is, as the black community maintains, based in part on the exploitation of blacks, then the acquisition of this same prosperity by blacks could well occur only if another sizable minority is exploited.

---

139. Jacquelyn Grant, "Black Theology and the Black Woman," in *Black Theology: A Documentary History*, pp. 421-423.
140. William Calamine, "Black Theology in Postmodern Configurations," *Theology Today*, April, 1993, p. 68.

At the Theology of the Americas Conference held in Detroit in 1975 most of the black theologians strongly defended the virtues of capitalism. Cone recalls and admits that despite their black rhetoric, black theologians wanted a healthy slice of the capitalist pie, rejecting Marxism in the process.

> "We lost our radicalism and became 'big time' executives and seminary professors."[141]

Black theologians have expressed admiration for and solidarity with Latin American theologians who have not only criticized social and economic structures but have offered concrete socioeconomic alternatives, not hesitating to borrow from "foreign" ideologies (Marxism being the most prominent) in their search for solutions to the problems they descry. However, such admiration has not led to a similar kind of historical and structural analysis essential to an understanding of where the black community is now and what is required to be done if it can have genuine hope for a better future. The reason why admiration has not led to emulation may be explained by the fact that black theology has some serious reservations about the completeness of Latin American theological solidarity with the black community. For one thing, the black theologian complains that Latin American theology has been slow to condemn the North American theology for its failure to attack the evil of racism in America, for not focusing on the rights of the black diaspora. But, considering the many problems that occupy the Latin American theologian in his own territory, where military governments murder their own people at alarming rates, the expectation is both unfair and unrealistic. The expressions of solidarity with the black community and the extrapolations which can be made from the liberation theological model seems, to me, to be all that one can expect. This all the more so since social analysis remains an important part of black theology's unfulfilled agenda. Cone regrets this did not happen during Martin Luther King. Jr's time.

> Although Martin and Malcolm began to consider classism as an evil along with and integral to racism, it was still difficult for them to incorporate this analysis into their thinking and activity....[142]

---

141. Cone, *For My People*, p. 94.
142. James Cone, Martin and Malcolm, p. 286.

And:

> As long as Martin and Malcolm remained within the first phase of their intellectual and political development, they excluded class analysis or minimized its significance and concentrated their energies on eliminating racism....[143]

Black Theology and Marxism

The black church and the various forms of Marxism have approached reality from different directions, but they do have a surprising number of objectives in common. Their perspectives are so unique and so far from mainstream America, that they share the experience of isolation. Both employ the same methodology of the dialectic, the unfolding of events in history, whether, as in black theology, the Holy Spirit is the prime mover or, in the case of Marxism, society is in constant transition, as old orders die and new ones evolve, depending on the constant shifting of socio-economic factors. For the black community the status quo is unacceptable; the transitional society is the only acceptable model for the future. The concerns of both communities, therefore, overlap, but both have fallen victim to survival by competition, a condition shared by minority communities everywhere. The daily struggle inhibits both long-term strategizing and the leisure of dialogue in a familiar setting. Capitalism thrives on the creation of marginalized peoples who claw at each other for their share of the "pie." Unless, by some miracle, racism ceases to be a significant factor in the assigning of status in American society, then some form of the socialist option is one that the black community might spend time considering, assuming it is willing to strategize in the political arena with like-minded whites. The acceptance of black talent and expertise in the market place is something to be jubilant about, but it does not negate the necessity for a radical reconstruction of society if the basic condition of the black community is to improve. The black church speaks of liberation, but has no agenda for its realization beyond partial inclusion in a society that is less than welcoming.

Black theologians and Marxist thinkers are strangers, but their concerns overlap; both focus on the plight of the oppressed, so they do have common ground on which to establish a serious dialogue. If they can avoid the provincialism of "conversion of the other," seeing in each other

143. Ibid, p. 282.

valuable and productive partners, both in the sphere of economics and race relations, they can make a contribution to all of society. There is plenty of room for maneuver; neither are monolithic bodies of thought. Black theology contains the seed for the resolution of racial and ethnic harmony; Marxism offers an alternative to the specific economic establishment that ravages minority communities.

Marxism Provides an Alternative View

Marxist thought contains both a theory of history and an understanding of capitalism that offers white and black scholars alike concepts that both challenge capitalism and suggest alternative ways of looking at economic institutions. In Marxism, black theologians find the principle that all human societies develop in transitional stages according to their systems of production. Within the mystery of the dialectical movements in history there lies the answer to the question: Will things ever change? The Marxist answer takes some of the mystery away by claiming that conflicts arise within the system of production as the mechanics of the system become outmoded, provoking, ultimately, social change. Man injects his own, perhaps revolutionary, activity to shape the social configuration and pace of this change.

What contribution can the black scholar make to this analysis? Since the black community, together with other minority communities, is directly impacted by the distribution patterns of the capitalist system, it is they who have the greatest incentive to explore, write and teach economic alternatives. Marxism, a theory that minimizes the role of religion and culture, is not the answer; it is, however, an important catalyst for further study, for refocusing some of the abundant energy residing in the black community, from racist to economic issues, from alienating attachments to material goods, to the liberating force of study and analysis. Marxism's tragic life in the Soviet system discourages us all from renewing its study but, at some point in time, we need to grow beyond that hesitation, and rediscover what it has to offer, at least in the way of provocative concepts. It is time to move on. All scholars, black and white, if they are true to their gifts, eschew the status quo in any area of human endeavor, and contribute their talents to man's ongoing historical journey. Truth is a journey, not an establishment fact.

Cornel West and the Dialogue

Cornel West has made the most significant contribution so far in attempting to provide a socioeconomic strategy for black theology. He points out that black theologians have adopted a dialectical methodology. They first negate white interpretations of the gospel message, go on to present their own perception of biblical truths, and then offer new perceptions of biblical texts. This theological process is akin to Marxist dialectical methodology which begins with exposing the fact that the misunderstanding of capitalist society by bourgeois-theoreticians has supported the exploitation of minorities in American society. It would seem then that black theologians and Marxist thinkers would be natural allies, frequently communicating with each other. The fact that this has not occurred is due, says West, to the fact that the dialectic which occurs in black theology occurs unconsciously, while with the Marxist it is a result of a conscious effort. West further states, however, that a dialectical methodology is implicit in black theology. The different histories of the black church and Marxism, as well as their different perspectives on the human condition accounts for their separateness, and separation carries with it the burden of misunderstanding. Both, interestingly enough, have been marginal in American society and have thus been preoccupied with their own survival, taking little notice of one another.

The failure of black theologians to analyze the manner in which the existing productive and social structures relate to the exploitation of blacks is, according to West, the reason why the rhetoric of black power has not been translated into productive and meaningful gains for the black community. Unlike Marxist thinkers, he says, the black community has failed to understand the internal dynamics of a society which is chronically exploitative of powerless minorities. There is the additional problem of confusing Marxist analysis with the Cuban or Bolshevik models of liberation, models that were designed to suit particular historical and cultural situations; their failure to develop strong economies in both cases certainly discourages further exploration. Each society must analyze its own unique socio-economic reality. All of us need to recall that 61 percent of the American population own only 7 percent of the nation's wealth, or that multinational corporations not only monopolize the market place, but are also subsidized in one form or other by the government In this context Marxist analysis can be useful since it emphasizes that true democracy includes the need for group participation in all levels of production. A theology that stresses the need for inclusion in ill-gotten wealth at the expense of a different oppressed community is itself in need of critical analysis in the light of the gospel message. Cooperation with the white community in this area of life can lead and contribute to a more meaningful inclusion, that of a common humanity. Once there is an understanding that the majority of whites also suffer various degrees of powerlessness, sharing low-paying jobs, inferior housing and tenuous heath care, a sense of solidarity can emerge, even if filtered through the thick maze of racist attitudes. Black Power brought black awareness; economic analysis can bring a sense of solidarity with a white community that bases part of its racism on the need in a capitalist society for a "bottom group."

Black Theology Can Contribute to Marxism

If Marxist analysis has a weak link, it is its lack of a comprehensive understanding of the role of culture and religion in man's history. And in no other theological model are these two important elements more explicitly and profoundly treated than in black theology. The role of religion in a people's resistance to oppression finds no better example than in black theological literature, and the Afro-American struggle for survival in an alien, alienating world. The black experience belies the Marxist assumption that racism is, in the final analysis, a factor of misplaced material interests. Marx himself, because of the clarity of his vision of the role of the institutional church in the society of his time, makes a false extrapolation when he presumes that religion is simply another instrument of social domination. He and his followers make the same mistake when they treat the question of culture, viewing it superficially as merely a reflection of class interests. Liberation theology, having benefited from the many penetrating insights of Marxist analysis, itself also suffers, to some extent, from the adoption of misguided Marxist extrapolations. James Cone and others have pointed this out in their writings.

Black theology, on the other hand, makes proper and valuable distinctions between popular culture that liberates and popular culture that prevents the light of truth from filtering through to the individual and social psyche. It is, therefore, important to all Americans that the black theologian remain close to the people in the street; active participation in the sophisticated life of the academic, a life far removed from the struggle for daily bread and dignity, can lead to a deadened sensitivity to gospel-related values. The intellectualization of Christian theological principles is, as with his colleagues in the white theological establishment, the black theologian's greatest temptation and danger. The same is true of the white theologian, but that is almost irrelevant because the white theological community has little to offer religious development in America, unless and until it comes to grip with America's greatest, most obvious sin. At this juncture in American history, the white theological community is sounding brass and tinkling symbol. Its lack of relevance to American religious life is directly related to its failure to seriously and incarnationally come to grips with racism. If this seems too sweeping a statement, compare it to Heidegger's claim that he is the first thinker in the entire history of philosophy who raised the question of the sense of Being. According to Heidegger, man has lived for centuries completely oblivious of the question of being, a negligence responsible for "the decline and crisis of man's history on this planet." The white community's anthropological explication of the gospel message is but another example of Heidegger's observation. The European preoccupation with the search for that-which-is becomes the blinders that prevent insights into Being; a condition with which black religion is not afflicted.

Jesus Said

The plastic Jesus who adorns the living rooms of the pious has become the standard throughout the churches, even amongst the educated. If some people reserve their religion to Sunday morning services, it is also true that the intellectual community reserves its talent and energies to that-which-is, leaving the exploration of Being to others. This has all sorts of practical consequences. With all the resources at their disposal, the complex of Universities manned by religious groups or ministries are nowhere to be seen when serious issues impact themselves on American society. Where was the church when Martin Luther King and his followers were beaten and hosed? Where was the church when the Vietnam agony spilled out into the streets of American cities from coast to coast? What does the church have to say about the pop culture, about Madonna, rap music? What voices from the choir are heard when American military might is used to shore up a President's declining popularity? The silence is deafening. The plastic Jesus, adorned in pretty dresses, looks with vacant stares at the strife-torn streets of Los Angeles, as our black brothers and sisters despair of humane inclusion in a society that their back and brawn have built, as many newly-arrived immigrants rise, mercurially, to the top rungs of the economic ladder. When they protest such glaring inequities they are accused of, yes, *racism*.

But Jesus continues talking. He speaks of the forces that are arrayed against the oppressed. He keeps shouting at us that if we want to be perfect we must share our wealth with the poor. From pulpits across America he keeps explaining, patiently, that it is easier for a camel to pass through the eye of a needle than to enter the Kingdom of Heaven. But we just say "Yes!" and go right on strategizing how to squeeze more from labor, from the marginalized, from the minority communities. In our supermarkets located in America's ghettoes, rotting vegetables are spruced-up and sold to the poor who cannot travel to the suburbs. People listening to Jesus ask the question: If you demand that we share our wealth, that we stop oppressing and exploiting the poor, then who can be saved? The question is an admission of collective guilt. The answer is cryptic. "What is impossible for man is possible for God." It's a consoling answer, we believe, but we may be wrong.

Maybe we think that this irritant will go away. After all, how serious could Jesus be? How serious is anyone taking Him? No one is enforcing a thing he says; even ministers who dare to preach his message nakedly, are punished, alienated from their brothers, forced into finding consolation amongst the pagans, denounced. Statements of Conscience, Pastoral Letters, Synod Declarations pour in a steady stream into church and University libraries, issued at public news conferences. But there is "an understanding:" don't *do anything* about the calls for racial harmony, economic justice, international exploitation. If you do, we'll *all* be in trouble with the establishment. If church authorities do not abandon the ministers who challenge the international corporations, who will support the charitable institutions of the church? No one!

Only the black, marginalized, church, in the person of its ministers and choirs, walk the streets, call Jesus' name, shout his message in the long and dangerous corridors of power. They are out there alone, while the powerful white church keeps one eye on the cross and another on the vagaries of Wall Street, one hand on the bible, the other entwined in the tentacles of mammon. And all the while, the plots against Jesus continue in the boardrooms. The chief priests and the teachers of the Law are still afraid of the people, so they carefully plan the death of their advocate.

They offer Judas a few pieces of silver (he is easily co-opted), pleased to rid themselves of so irritating a problem for a fistful of coins.

Economic Component of Racism

The challenge to a new generation of black theologians consists in the reality that now, more than in the last fifty years, the focus on matters economic will prove very difficult. The global presence of the Eastern Bloc of nations, inviting oppressed peoples to the socialist paradise, has all but disappeared. Also, the Bloc's miserable economic track record shall linger on the West's collective memory for generations, making it almost impossible to suggest an alternative to the present economic system. Given this new situation, it is not encouraging that the black theological community has limited its understanding of participatory democracy to the desire to share in white affluence. Such a desire begets the danger of becoming a part of the very system attacked as exploitative, generating a theological and logical disconnect. The relationship between racism and economic exploitation needs further study. It is clear that a comprehensive social theory is needed before the hopes raised by the black theological community can become a reality.

To the capitalist victor go the intellectual and emotional spoils. Only a fresh new crop of inspired black theologians would be young enough and energetic enough to take on this task. The white community, so long as it can "look down" at a less affluent sector of the population will almost certainly prefer to support the status quo, believing, in an illusion, that it is participating in the good life. Only the black theologian (and his constituents) has the existential background needed to move America forward. A new focus with the same theme, "We Shall Overcome" is the next contribution black theology will make to the Republic. We need to pray that black America will produce such sisters and brothers in the theological community, for they represent our salvation. The black academic community, if it produces more persons the like of Cornel West and Jaime Phelps, can make a great contribution toward fulfilling the American dream for all of us. While I am no advocate of the "suffering servant" theory, I do believe that only black America has the moral fibre needed to save us from our collective sins.

The Need for Black History

The disruption of African family life and the loss of historical identity that flowed from the complex apparatus of slavery has produced an entire people who have no history, in the usual connotation of that word. Current history, as it is lived in the streets of America's cities needs to be evaluated. It is clear that division within the community leads to weakness vis-a-vis the white community and, more recently, other minority communities. There is a need for coordination of effort because a minority that differs in racial composition with the majority cannot seriously contemplate "taking over." Confrontation must be limited in scope, have specific targets, and back-up positions should things go wrong. The struggle requires discipline, study, sophistication. Black unity also demands a cessation of violence against brothers and sisters, an energy wasted, abused. It is as wrong, strategically, to act in the streets as if justice already exists, as it is right to demand that justice be done. Leone Bennet points out that the struggle is still in a period of transition:

> It is too soon to say where all this will end. Some forces in the black community are still developing, other forces are working at cross purposes and conditions are not yet ripe for the Black Rebellion *to become what it is.* This is a transition period. The Rebelling is seeking its truth, its laws, its life. It is growing, groping, feeding on itself, adapting itself to the action and lack of action of white America. It is inventing itself by countermove in the white darkness. [144]

The Historical Non-Record

The Black community continues to grope for identity. We are reminded of this every time a black "Johnson," "Ryan," "Carmichael," or "Jackson" makes the news in sports, politics or religion; the African connection is not identified in the name. As an avid follower of Notre Dame football, I have often wanted to write the University's student newspaper asking them to reflect on their reports of "Irish" victories on the field as, these days, young black athletes bring new glory and praise to the school's football tradition. Is not such reporting a contribution to what James Baldwin describes in his writings as the facelessness of the black person in American society? Does not the "Catholic" sense impel us to make these seemingly small but significant changes in our reporting? Obviously it has not,

144. Leone Bennett Jr., *The Challenge of Blackness* (Johnson Publishing, Chicago, 1972), p. 29.

and therein lies a story of the white community's lack of self-understanding as "church."

While working in the Caribbean, I searched for historical records written from the black perspective. In Grenada *the* historical record is that of a white missionary, Father Demas, a man who loved Grenada and its people, but who never "lived" the black experience in that lovely spice-isle. Of the many crimes associated with slavery, the missing links to the black man's past is perhaps the most heinous. We need to know where we come from in order to chart a course into the future; our fascination with history derives from the fact that we *are* "historical." History is the umbilical cord binding us to our genetic past. The tenuous existence of the slave and the prohibition of his existence *as African* developed in his generation a preoccupation with *today* and this was passed on to succeeding generations. There was no certainty about tomorrow, and even if there were, abuse and humiliation was in store. "Tomorrow" has had little interest for a community that does not find there growth, new challenges, and the construction of a humane reality. "Tomorrow" conjures up images of dull, unchallenging, and unproductive survival. Man lives by the self-portrait he draws in his imagination, a self-portrait of his historical existence. Because of the lack of data, this task for the black community is as difficult as it is necessary. Black Power, an enduring moment in American black history, served as a very useful vehicle for the illumination of black pride and the articulation of black protest. Black theology and black scholarship provide other (but not isolated), necessary moments. The perception in the white community is of a black community that lives only for the moment; what is lacking in that perception is an understanding of the white community's decisive role in creating that condition. The white community has both the power and the resources to provide a reason for "tomorrow;" as yet, it lacks both the will and a proper understanding of Christianity to do it.

A House Built on Lies

The white community has constructed a false American white superiority by declaring as a fact a non-existent Black inferiority. Cleage points out that from the time Black people came to America as slaves, Black inferiority was the framework within which they were forced to live their existence. From the slave ship, to the plantation, to the hangman's noose, the

Black man's inferiority was "proven." Separated from those who spoke his language, living in an alienating land and required to accept a foreigner's definition of his identity, he lived only not to die. He had no other, independent, existence. His illusory hope and dream is that *he*, as an individual, will be exempted from the general rule of slavery, a hope based on the occasional exception to the rule. Integration is no longer a word that lies easily on his tongue. Too much has happened, too many promises broken. No longer is the Black man willing to be defined by the white man; he has begun, after a very slow start, the process of his own definition, the long-delayed search for his own identity. Not an easy task for the slave who became "free," and saw his future in racial integration. His own prophets held out the hope that this would happen. Martin Luther King Jr. built his message on it, and received a Nobel Peace Prize, not for succeeding, but for trying.

The social choices have become fewer, more radical. The goal ahead is Black cooperation, unity, study, scholarship, high morality, self-discipline., a return to the survival techniques of pre-Bellum days. The absorption of the corruption rampant in white society has devastated the black community. It needs to regroup, as a group. Cleage quotes the Black Christian Nationalist Creed:

> I Believe that both my survival and my salvation depend upon my willingness to reject INDIVIDUALISM and so I commit my life to the Liberation Struggle of Black people and accept the values, ethics, morals, and program of the Black Nation defined by that struggle, and taught by the Black Christian Nationalist Movement.[145]

Even the black nationalist, the most radical of the radicals, is faced with an almost insurmountable choice. In spite of all that has happened, all the misery, the neglect, the humiliations, America is home to the black community. The black music, life-style, sports figures, and poets; it would not be America without that contribution—without Whitney Houston, Cornel West, Magic Johnson, Langston Hughes, Michael Jordan, Arthur Ashe, Frederick Douglas, Jackie Robinson, Colin Powell, Ella Fitzgerald, Nat Turner, and Michael Jackson. They lift us all on splendid wings; we admire (envy?) their courage and dignity, their strength. It would not be America ever again without them.

---

145. Cleage, p. xii

Shamefully, look at the awful choice we put before these beautiful brothers and sisters. We force them to build a Black value system, a separate Black nation right in the midst of our largest cities, and even in the countryside. We force Blacks to reject our values, and the values of the white world everywhere. Then we say: "They can't live with us." We alienate them, and then accuse them of having "a chip on their shoulder," of rebelliousness. "They reject our values," we grumble. What values? Our *Christian* values? They're not *real* Americans, shout second-generation Europeans and Asians. We structure the economic system so that "they" can't find jobs, and we accuse them of being lazy. We say, "if you don't have a job, go on welfare," stop complaining. Then we complain because they are on welfare. We scream our heads off about the evil of abortion; then we accuse black women of being baby-producing machines. And the black church, the cultural and religious center of the black community takes up the cause, the call to arms, because it must, because it has the heavy burden of telling its people the truth, of leading the fight. They don't really want to do that. They would rather be havens of peace, centers of conflict-resolution, but don't have the luxury of the white churches, keeping everybody happy, staying above the fray. The white hierarchy is not required to march in the streets, to join the demonstrations, to sacrifice to bring about social change. It would be out of character. Why would the establishment march against itself? It won't; it can't, period.

The black ministers get their hands dirty, mix with the crowd, and then are accused of mixing politics and religion. The white church has mixed religion and politics since before the American revolution, since before there was an America, since Constantine. It is, in a way that no other social institution is or can be, *the* political establishment. Who dares defy them? Presidents and Governors court their favor. It has all this power, all this influence, but this power and this influence is used to maintain power and influence. It's called "preserving the institution." Of course, the word "church" is used here to mean the official institution, not the countless numbers of faithful, clergy, nuns, church workers in countless numbers who *do* love, who *do* march, who *do* put their "neck" on the line. You meet them everywhere, everywhere that is except in the chanceries; and even there one meets an occasional maverick. They hope to stem the tide, to make the Black nation unnecessary, to make America so welcoming to its own, that the black lamb can lie down with the white lion. The hope springs eternal, motivated by faith in the gospel message, in the personality and person of Jesus Christ, in the constitutions of their churches, their religious orders, the presbyters.

For now, however, the enforced alienation of the black man continues, and it is very painful. He doesn't *want* to build his own nation, establish his own values; he is proud of the sports figures who contribute to making America the sports center of the world. He really *wants* to be proud to be an American. He wants to feel the thrill of "God Bless America," and he does, but with guilt feelings, because he has been forced to build another nation, right here, where he has lived and worked and died longer than any other group of peoples, excepting the Pilgrims and the Indians.

Implications for Jeffersonian America

As a student at William and Mary College in 1760, Thomas Jefferson was tutored by Dr. William Small (who drew up the agreement between the Scotsman, James Wyatt and the Englishman, Matthew Boulton, early steam engine pioneers). Dr. Small, much to our good fortune, introduced Jefferson to George Wythe, law professor at the College. From him Jefferson leaned the art of carefully and with great exactness, yet intellectual flexibility, studying the law. The democratic quality of Jefferson's mind is imbedded in his letters, documents and books, and setting aside a space for their study is appropriate for every American, especially those who hope to devote their lives to public service. Wordsworth, writing in *Lyrical Ballads*, in 1798, complained that eighteenth century poetry and prose was decadent and outmoded, the wounded style of an old and dying generation. Fifty years before, however, one American, Jefferson, had already moved beyond that style, writing with the simplicity of the countryside around him and the fresh approach of a new-world gentleman. He remained untouched by the legal style of his time, calling it "lawyerish." A distinguished gentleman, whose lineage was derived from the Old World, he set the example for a new age, a new world of literature, and politics. Of the Declaration of Independence he observed, in a letter to Henry Lee, that it did not aim at originality of principle, nor was it copied from any particular previous writing, but was simply intended as "an expression of the American mind."[146]

This scholarly mind, this dignified gentleman declared, through the language of the Declaration, that revolution is an accepted method or mechanism of government. "Whenever any form of government becomes destructive of..." etc., it is the right of the people to alter or abolish it. Like Locke, his soul-mentor, he was very serious about the right of revolution, saying that the tree of liberty needs to be watered by the blood of patriots from time to time. He even threatened revolution against the American government when upset over the Kentucky Resolutions of 1798 against the Alien and Sedition Acts.

Based as it was on such sophisticated men of simplicity, American political philosophy has been uniquely absent of theoretical writings. *The Federalist* is about the only major work of American political philosophy. Whatever political genius America possesses is found in public documents, legal decisions, and political debates. The implications of all of this for the American problem of racism is that, collectively, we can make a decision to stop *it*, to start anew, to breathe new life into the body politic. Our founding Fathers wanted it that way; we need to go back to our roots.

Jefferson Said

If racism has been a long night of terror for the black community, it has been an equally long descent into moral inferiority for the white community. Ironically, but in keeping with the clear message of the gospel, it has made the black man, though wounded, strong, and the white man weak. The white man philosophizes and takes his orders from the grave, because his message no longer refers to the living. Jefferson, in a letter to James Madison, asked the question, whether one generation of men has the right to bind another. His answer is that the earth and the use thereof belongs to the living; the dead have neither the right nor the power to interfere. If that be the case, why are we still living in the intellectual carcasses of bodies long since buried and corrupted?

There was also the question of who owns the property of the dead. Jefferson answers that question by stating that a man does not have the right to tie up, apart from or in opposition to the legal framework of a society, all future use of land, once he has departed; it is the society's needs, not the wishes of a dead man that really matter. In a brilliant analysis of the nature of the living person, Jefferson states that one generation "is to another as one independent nation to another." He argues that on

---

146. Letter of May 8, 1825, to Henry Lee, from *The Writings of Thomas Jefferson (New York, 1898-1899), Vol. X, p. 543.*

similar grounds no society can make, in perpetuity, either a constitution nor even a single law. He suggests that the life of any constitution be ended after nineteen years, claiming that even if this is not expressly stated in the nation's laws, it becomes a reality by force of the dialectical nature of nature.

America needs, not a resurrection, but a new birth. We are living with ancient prejudices and massive inequities; our black brothers and sisters suffer, and experience not only benign neglect but the social awareness of the leper. If the fruit of one's labor truly reverted to the self, then the land belongs to the black community, and we in the white community occupy it only at their sufferance. The major contribution the white community has made to the southern plantation is its sophisticated ability to exploit it and expend its capital. It is time we repented for our sins; it is time we stopped hiding behind dead men's bones; it is time for America to come alive again.

Jefferson comments, in his First Inaugural Address of 1801, give us both the motivation and the philosophy to eradicate this cancer eating away at the heart of America:

> A wise and frugal Government, which shall restrain men from injuring one another, shall leave them otherwise free to regulate their own pursuits of industry and improvement, and shall not take from the mouth of labor the bread it has earned. This is the sum of good government, and this is necessary to close the circle of our felicities.

CONCLUSION

Before the publication of books and articles on black theology, black people and their concerns were not a noticeably visible part of white American theological deliberations. Not even the presence of Martin Luther King, Jr., and the civil rights movement stirred the white theological community to enter into dialogue with black history and experience in the search for a Christian resolution to the problem of American racism. Although it is generally recognized that racism is America's most serious moral dilemma, the attention given to this problem in white theological seminaries remains minimal. The failure of the white professional theologian to respond to the black struggle for justice convinced black intellectuals that a "black" theology was needed if the black community was ever to enlist the forces of religion to its just cause. The white clerical voice perpetuated the theme of slavery:

> If you are black you can be poverty-stricken, you can be brutalized, and you can still be saved. Your children can be discriminated against and denied a decent education and you can still be saved. You can live in a neighborhood from which all the decencies of life have been taken and you can still be saved. This kind of primitive Christianity the Black slaves received from their white slave masters.[147]

Black theology has made remarkable progress in a short time in spite of many external and internal problems. It had to overcome a natural caution toward its efforts on the part of a sophisticated theological community with a long history of academic development. It had to convince scholars both in and out of the field of theology that passion and rhetorical language were admissible in what has been a field characterized by careful, judicious and moderate language. Also, the fact that black theology as an academic discipline has not been widely accepted by the black church itself underscores its need to develop a consistent strategy in the black community. Its failure to produce a viable analysis and answer to black

---

147. Albert B. Cleage Jr., *Black Christian Nationalism* (Luxor Publishers, Detroit, 1987) Introduction, p. xxix.

poverty has made its credibility amongst black congregations problematic. The black theological agenda, it seems to me, should remain open to new ideas, new alliances and more thoughtful analysis.

Black theology had to deal with the suspicions and reservations of white theologians who tried to sort out what they perceived as an intermingling of theology and political activism. The use of revolutionary language and concepts within the context of academic theological debate did not make the meaning of black theological discourse easy to communicate. These suspicions and reservations persist today, while the language of Latin American liberation theology has become familiar, and even dominates international theological discourse. Of great importance is the need for black theology to remain faithful to the black church tradition from which it arose, and provide a sociological analysis of the American economic system, which it claims continues to deprive the black community of its rightful participation in the benefits of living in American society. Although black theology, as an academic discipline, is relatively new on the American scene, what it lacks in full theological development, it more than makes up for with its cogency, righteousness and power.

Given these problems and limitations, it is clear that black theology has creatively established itself as a theology to be recognized and commented on. It has prompted serious reflection in the American theological community on the nature of Christian America and the depth of its understanding of the Gospel message. It has proven itself to be, in its clear and unequivocal language, superior to the morally-hesitant theology practiced in the white community. Its prophetic voice will continue to bolster theological pluralism in America, and bring to the attention of white Christian America that unless the problem of racism in our society is resolved, the American dream will continue to elude us.